MW00837537

Springer Series in Optical Sciences

Volume 241

Springer Series in Optical Sciences is led by Editor-in-Chief William T. Rhodes, Florida Atlantic University, USA, and provides an expanding selection of research monographs in all major areas of optics:

- lasers and quantum optics
- ultrafast phenomena
- optical spectroscopy techniques
- optoelectronics
- information optics
- applied laser technology
- industrial applications and
- other topics of contemporary interest.

With this broad coverage of topics the series is useful to research scientists and engineers who need up-to-date reference books.

Kan Yao • Yuebing Zheng

Nanophotonics and Machine Learning

Concepts, Fundamentals, and Applications

Springer

Kan Yao
Walker Dept of Mechanical Engineering
Texas Materials Institute
The University of Texas at Austin
Austin, TX, USA

Yuebing Zheng
Walker Dept of Mechanical Engineering
Texas Materials Institute
The University of Texas at Austin
Austin, TX, USA

ISSN 0342-4111 ISSN 1556-1534 (electronic)
Springer Series in Optical Sciences
ISBN 978-3-031-20472-2 ISBN 978-3-031-20473-9 (eBook)
https://doi.org/10.1007/978-3-031-20473-9

This Springer imprint is published by the registered company Springer Nature Switzerland AG
The registered company address is: Gewerbestrasse 11, 6330 Cham, Switzerland

For my parents, who have been understanding and supported me through my life.

– Kan Yao

To Yanhua, Steven and Mia.

– Yuebing Zheng

Preface

The study of interactions between light and materials has a long history, dating back to perhaps as early as the time even when the nature of light had not been settled. As science keeps advancing, there is one line in this study that can be traced by looking at the decreasing dimensions of the materials, from optics to photonics, and all the way down to nanophotonics. Nanophotonics studies light-matter interactions at the nanoscale, where the materials in most cases are structured into subwavelength building blocks so that exotic optical properties beyond those of bulky materials emerge. Over the past two decades, nanophotonics has attracted rapidly growing interest and become a vibrant research field that contains both fundamental and application-driven studies. Depending on the materials, geometries, sizes, and arrangements of the constituent elements, nanophotonics can be categorized into several subfields, including plasmonics, metamaterials and metasurfaces, photonic crystals, photonic-integrated circuits, and other resonant nanostructures that can perform photonic functions. These devices operate on different mechanisms, enabling unprecedented opportunities to control light at the nanoscale for unveiling new physics and achieving fascinating applications not possible with conventional techniques.

Artificial intelligence (AI), seemingly on a totally different subject to nanophotonics at a first glance, is currently among the most promising techniques that can revolutionize the world from many aspects. The history of AI is a bit less than 80 years, with the beginning marked by the early research on neural networks in the 1940s. The popularity of AI nowadays has gone far beyond computer science and infiltrated many other research fields, such as physics, chemistry, materials science, and biomedicine, to name a few. After the astonishing success of the computer program AlphaGo from Deepmind defeating the top professional Go players, even the public might develop the idea that a new era, where AI is competitive with human intelligence in completing certain tasks (i.e., weak AI), has come. In science and engineering, especially those fields fitting well with big data, the expected tasks include materials discovery, drug creation, and so forth.

It is interesting to picture how AI would transform nanophotonics. Although surely not a panacea for all the remaining challenges, AI can potentially assist the design of nanophotonic devices. Conventional inverse design is based on a trial-and-error process, which can be extremely labor-intensive. The initial guess of the solution to a design task usually relies on human knowledge, a combination of intuition, the physical insights revealed from the study of modal systems, the experience accumulated during the previous practice, and reasoning. It is then examined with simulations by solving Maxwell's equations but unlikely to meet the desired performance in one shot. Therefore, adjustments to a handful of parameters and re-evaluation of the new designs need to be repeated until some preset criteria are reached. A variety of optimization methods have been developed to prevent this process from being almost blind. They also make it possible to search the enormous design space in a more comprehensive manner, yielding complex and non-intuitive structures that cannot be parameterized. But still, the conventional workflow requires considerable computation power and time for every design task, which could explode as the complexity of the devices and the scale of integration increase. AI or, more specifically, machine learning provides new solutions that work in a totally different logic. Through training, these so-called "data-driven" methods leverage many instances of known devices to improve the ability of finding optimized designs for a certain set of design tasks. The questions on whether and how AI will benefit inverse design remain fairly open. On the one hand, no clear evidence has suggested that the efforts in generating sufficiently large training sets for AI programs can be less fierce than those in the trial-and-error and optimization process. On the other hand, there certainly exist AI-related techniques, both algorithms and hardware, that can change the game in some way. In a nutshell, the application of AI in nanophotonics, including but not limited to inverse design, appears worthwhile and deserves more research efforts.

What makes the combination of AI and nanophotonics more interesting is the other side of the coin. With the explosive growth of machine learning in recent years, the computing hardware based on general-purpose processors becomes inefficient in implementing neural networks, raising the pressing need to develop application-specific hardware. Compared with the solutions based on electronic architectures, photonic circuits that can process coherent light signals are superior in speed and power efficiency. Some recent advances have demonstrated that specially designed nanophotonic circuits or structures can perform machine learning tasks like inference. Therefore, nanophotonics is not just fueled by AI passively; it offers improvement in return, making their relationships interactive.

Because the backgrounds of these two fields are very different, there is often a knowledge gap for people interested in this topic from either side. The goal of this book is thus to introduce the basics of nanophotonics and machine learning, especially deep learning, and to help the reader to get some sense on how they work and can be utilized to enhance each other. The first two chapters brief the fundamentals of nanophotonics, starting from surface plasmons and Mie resonances in modal systems and ending with representative nanophotonic devices and platforms. Although plasmonics and metamaterials have been discussed by several classic books separately, some emerging concepts, such as metasurfaces and dielectric photonics, might not have been covered adequately. As the preparation for the later part on applications, these two chapters are intended to provide the essential knowledge to tour the field of nanophotonics. Chapter 3 is analogous to the first two but with the subject shifted to the fundamentals of machine learning. The very basic as well as most popular models and algorithms are discussed to close the first part for knowledge preparation. Chapters 4, 5, and 6 present selected examples to illustrate how the concepts can be put into effect. Whereas Chaps. 4 and 5 both discuss the application of deep learning in nanophotonics, the former focuses on inverse design, and the latter spans more diverse usage. Lastly, Chap. 6 flips the coin to introduce how machine learning can be performed on nanophotonic platforms.

In order to serve the aforesaid goal within a compact volume while keeping balances between fundamentals and applications as well as between nanophotonics and machine learning, we choose to skip some introductory contents such as the basics of electromagnetism (for nanophotonics) and relevant math (for machine learning), which are widely accessible from resources elsewhere. Some areas of nanophotonics like photonic crystals and circuits are not included either, partly because they have been described in classic textbooks and monographs, and partly because they are on the margin of nanophotonics if one defines the latter by the subwavelength dimensions of the building blocks and their separations. With these in mind, we hope that the general reader of all levels, (under)graduate students, professionals, and researchers who are new to or have been working on either side of this interdisciplinary area will still find this book friendly and useful.

We are sincerely grateful to our colleagues and many others who have contributed to the completion of this book. In particular, we thank Mr. Rohit Unni for drafting an initial version of Chap. 3, and we are deeply indebted to Prof. Mingyuan Zhou and Mr. Xizewen Han for proofreading this chapter and providing invaluable suggestions in the midst of their busy schedule. In finalizing the book, requests for figures were made to some authors of the cited works. We wish to express our gratitude to all of them for generously sharing the original files or kindly directing us to the right person. Last but not least, given the extensive breadth of two very different domains of nanophotonics and machine learning and the rapid development of the playground where they interact, deciding the most suitable examples and references on the selected topics has been difficult. The current choice is made to the best of our knowledge. We apologize for unintendedly missing or slighting any important studies. Comments and suggestions are very welcome and will be greatly appreciated.

Austin, TX, USA

Kan Yao
Yuebing Zheng

Contents

Chapter 1
Fundamentals of Nanophotonics

Abstract The study of nanophotonics is centered around the interactions between light and nanostructured metallic and/or dielectric architectures (Koenderink et al, Science 348(6234):516–521, 2015). Usually associated with resonant behaviors, light at these nanostructures can be controlled in sophisticated ways not possible to achieve with materials in the bulky form. Therefore, before diving into the discussion of any nanophotonic devices and their design, it is beneficial for beginners to first establish an understanding of how light can interact with the basic building blocks of nanophotonic structures. This chapter is designated for this purpose. The two sections in the following focus on the fundamentals of plasmonics based on metallic structures and on Mie theory for dielectric nanoparticles, respectively.

1.1 Plasmonics and Surface Plasmons

Plasmonics, which studies the optical responses of and optical phenomena from metallic structures, is a major subfield of nanophotonics [1–4]. With its central concept, surface plasmons, being considered as a bridge between photons and conduction electrons in metals, plasmonics offers the opportunities for light propagation and localization at extended surfaces of metallic films and in bounded geometries such as metallic nanoparticles. In this section, we introduce the basic theory of plasmonics.

1.1.1 Surface Plasmon Polaritons

From the classical electromagnetic point of view, surface plasmon polaritons (SPPs) are waves that propagate along the interface between a dielectric and a metal. Because of their resonant interaction with the collective oscillations of free

The original version of this chapter was revised. The correction to this chapter is available at https://doi.org/10.1007/978-3-031-20473-9_7

K. Yao, Y. Zheng, *Nanophotonics and Machine Learning*, Springer Series in Optical Sciences 241, https://doi.org/10.1007/978-3-031-20473-9_1

electrons in the metal, these electromagnetic waves are essentially *confined* at the metal-dielectric interface with *enhanced* field intensity and exponentially decay along the surface normal into each medium. To see how these characteristics arise and to get more insights into other physical properties of SPPs, it is necessary to solve Maxwell's equations in this specific situation.

We first illustrate the simplest geometry on which SPPs propagate. Figure 1.1 shows the side view of a two-dimensional (2D) platform, where a flat interface coincident with the x-y plane separates two isotropic, homogeneous, semi-infinite media. For simplicity, let us consider the upper half-space ($z > 0$) is occupied by a dielectric with nondispersive permittivity $\varepsilon_d > 0$, the lower half-space ($z < 0$) is filled with a metal characterized by a complex, frequency-dependent dielectric function $\varepsilon_m(\omega) = \varepsilon_1(\omega) + i\varepsilon_2(\omega)$ where $\varepsilon_1(\omega) < 0$, and the electromagnetic waves propagate in the $+x$ direction. By assuming the fields are time-harmonic and have a factor $e^{-i\omega t}$, the governing wave equations in each medium are given by

$$\nabla^2\mathbf{E} + k_0^2\varepsilon\mathbf{E} = 0,\quad \nabla^2\mathbf{H} + k_0^2\varepsilon\mathbf{H} = 0, \tag{1.1}$$

where **E** and **H** are the electric and magnetic fields, respectively, $k_0 = \omega/c$ is the wavenumber with c the speed of light in a vacuum, and ε is the corresponding permittivity as defined above. For the system under study, the solutions can be classified into two groups depending on the field configuration. With the surface normal and propagation direction defining the plane of incidence, the solution that has the magnetic field perpendicular to this plane is designated as the transverse magnetic (TM) or p-polarized wave, and the other solution that has the electric field perpendicular to the plane is the transverse electric (TE) or s-polarized wave [5].

Let us examine the suitability of the two solutions to SPPs. In the case of TM waves, the nonzero field components are as follows:

$$H_y^{(d)} = Ae^{ik_x x}e^{-k_z^{(d)}z}, \tag{1.2a}$$

$$E_x^{(d)} = iA\frac{k_z^{(d)}}{\omega\varepsilon_0\varepsilon_d}e^{ik_x x}e^{-k_z^{(d)}z}, \tag{1.2b}$$

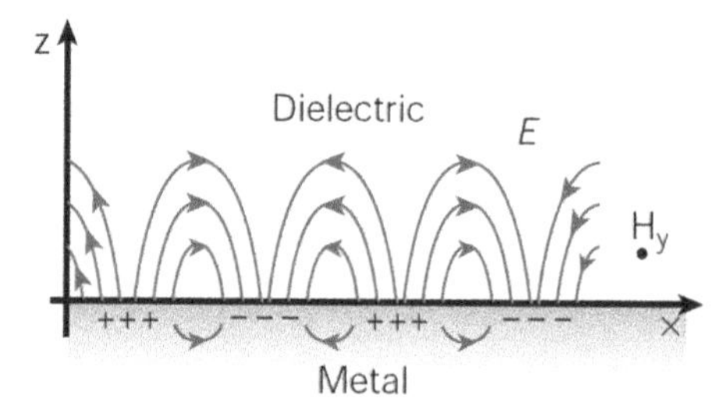

Fig. 1.1 Propagation of SPPs at a flat interface between a semi-infinite dielectric and a semi-infinite metal. Electromagnetic fields are confined near the surface and decay exponentially into each medium along the surface normal. The only allowed field configuration has the magnetic field component transverse to the propagation direction and surface normal. (Reprinted from [1] with permission of Springer Nature)

$$E_z^{(d)} = -A \frac{k_x}{\omega \varepsilon_0 \varepsilon_d} e^{ik_x x} e^{-k_z^{(d)} z} \tag{1.2c}$$

for the region $z > 0$ in the dielectric, and

$$H_y^{(m)} = B e^{ik_x x} e^{k_z^{(m)} z}, \tag{1.3a}$$

$$E_x^{(m)} = -iB \frac{k_z^{(m)}}{\omega \varepsilon_0 \varepsilon_m} e^{ik_x x} e^{k_z^{(m)} z}, \tag{1.3b}$$

$$E_z^{(m)} = -B \frac{k_x}{\omega \varepsilon_0 \varepsilon_m} e^{ik_x x} e^{k_z^{(m)} z} \tag{1.3c}$$

for the region $z < 0$ in the metal. Here, A and B are the amplitudes of the magnetic field in dielectric and in metal, respectively, k_x is the propagation constant, and k_z is the component of the wavevector in the direction along surface normal and is connected to k_x by

$$k_z^{(d)} = \sqrt{k_x^2 - \varepsilon_d k_0^2}, \tag{1.4a}$$

$$k_z^{(m)} = \sqrt{k_x^2 - \varepsilon_m k_0^2}. \tag{1.4b}$$

At the interface $z = 0$, the boundary conditions require continuity of the tangential magnetic field H_y, simply resulting in $A = B$, and continuity of the tangential electric field E_x, yielding

$$\frac{k_z^{(d)}}{k_z^{(m)}} = -\frac{\varepsilon_d}{\varepsilon_m}. \tag{1.5}$$

Keeping in mind that $\varepsilon_d > 0$ and the real part of ε_m is negative, Eq. (1.5) implies positive signs for the real parts of $k_z^{(d)}$ and $k_z^{(m)}$, which correspond to attenuation of the electromagnetic field from the interface into each medium, a configuration fulfilling the requirements of a surface wave. The negative signs are opted out because of their correspondence to the growth of field magnitudes, which is physically impossible in passive media. From now on, we rename the propagation constant k_x to the wavenumber of SPPs k_{sp}.

In Eq. (1.5), the wavenumber k_{sp} is related to the frequency ω in an implicit way. The explicit expression can be acquired by inserting Eq. (1.4) into Eq. (1.5) and rearranging the terms. This treatment leads to

$$k_{\mathrm{sp}} = \frac{\omega}{c}\sqrt{\frac{\varepsilon_d \varepsilon_m}{\varepsilon_d + \varepsilon_m}}, \tag{1.6}$$

the dispersion relation of SPPs that propagate along a flat interface between a semi-infinite dielectric and a semi-infinite metal. While experimentally measured dielectric functions can be used to calculate the dispersion in practice, it is instructive to study a simplified model where the property of metals follows the free-electron description

$$\varepsilon_m(\omega) = 1 - \frac{\omega_{\mathrm{p}}^2}{\omega^2}, \tag{1.7}$$

with ω_{p} the plasma frequency of a metal and the damping of electrons neglected. The value of ω should be taken in the range $\omega < \omega_{\mathrm{p}}$. Figure 1.2a illustrates the dispersion diagram of an SPP propagating along the interface between a dielectric ($\varepsilon_d = 2.25$) and a metal described by Eq. (1.7). For comparison, also shown is the dispersion of electromagnetic waves propagating in the dielectric, known as the light line. It is clear from the plot that the dispersion curve of SPP lies to the right side of the light line in the constituent dielectric medium. At low frequencies, the two curves almost coincide, meaning that an SPP acts like a TM-polarized wave incident at a glancing angle. As the frequency increases, the dispersion curve of SPP is gradually bent towards the right, showing continuous increase in wavenumber. Therefore, SPPs are bound at the metal-dielectric interface. They cannot radiate light or propagate into the dielectric medium, and, reciprocally, cannot be excited by simple illumination from the dielectric constituting the interface. Without having

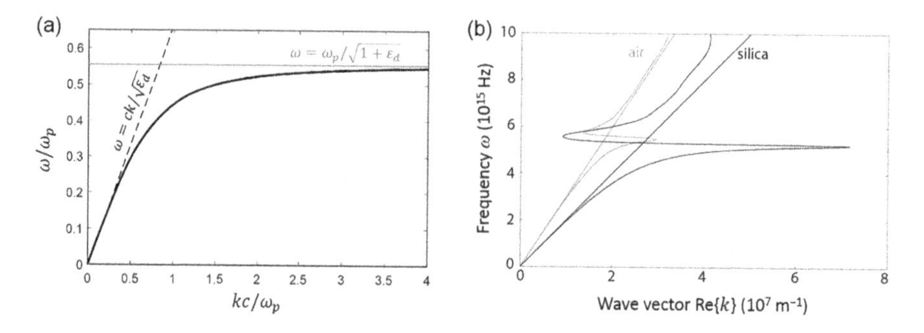

Fig. 1.2 (**a**) Dispersion diagram of an SPP at the interface between a semi-infinite dielectric (ε_d = 2.25) and a semi-infinite metal modeled as an undamped free-electron plasma ($\omega_{\mathrm{p}} = 11.9989 \times 10^{15}$ Hz for silver [4]). The dashed line is the dispersion of electromagnetic waves in the dielectric medium. The horizontal line indicates the surface plasmon frequency. Both frequency ω and wavenumber k_{sp} are normalized to the plasma frequency ω_{p}. (**b**) Dispersion of SPPs at an interface between silica and silver (black curve) and between air and silver (gray curve), along with light lines. The dielectric function of silver is taken from experimental data [6]. (Panel (b) is reprinted from [3] with permission of Springer Nature)

the damping of electrons taken into account, the wavenumber k_{sp} goes infinitely large under the condition:

$$\varepsilon_d + \varepsilon_m(\omega) = 0, \tag{1.8}$$

which is fairly obvious according to Eq. (1.6). This condition can also be obtained if we treat the SPP with infinite wavenumber, termed as the surface plasmon [3, 4], as an electrostatic wave and solve the Laplace equation for an electric potential function. The surface plasmon frequency ω_{sp} can be easily determined from Eqs. (1.7) and (1.8) as

$$\omega_{sp} = \frac{\omega_p}{\sqrt{1+\varepsilon_d}}. \tag{1.9}$$

With real metals, the wavenumber of SPPs can never reach infinity owing to losses. According to Eq. (1.6), k_{sp} will be complex-valued, and its imaginary part determines the propagation length of the SPP [7]:

$$L_{sp} = \frac{1}{2\mathrm{Im}\{k_{sp}\}}. \tag{1.10}$$

On propagating over the distance L_{sp}, which is about a few tens of micrometers for noble metals at visible wavelengths, an SPP decays to $1/e$ of its original intensity. The dispersion curve is folded at a finite maximum, as shown in Fig. 1.2b.

The solution of TE waves can be derived in the same manner, for which the non-zero field components are the following:

$$E_y^{(d)} = Ae^{ik_x x}e^{-k_z^{(d)} z}, \tag{1.11a}$$

$$H_x^{(d)} = -iA\frac{k_z^{(d)}}{\omega\mu_0}e^{ik_x x}e^{-k_z^{(d)} z}, \tag{1.11b}$$

$$H_z^{(d)} = A\frac{k_x}{\omega\mu_0}e^{ik_x x}e^{-k_z^{(d)} z} \tag{1.11c}$$

for the region $z > 0$ in the dielectric, and

$$E_y^{(m)} = Be^{ik_x x}e^{k_z^{(m)} z}, \tag{1.12a}$$

$$H_x^{(m)} = iB\frac{k_z^{(m)}}{\omega\mu_0}e^{ik_x x}e^{k_z^{(m)} z}, \tag{1.12b}$$

$$H_z^{(m)} = B \frac{k_x}{\omega \mu_0} e^{ik_x x} e^{k_z^{(m)} z} \tag{1.12c}$$

for the region $z < 0$ in the metal. Again, by applying the boundary conditions at $z = 0$, the continuity requirements of the tangential field components yield $A = B$ and

$$A\left(k_z^{(d)} + k_z^{(m)}\right) = 0. \tag{1.13}$$

As discussed above, in order to describe a surface wave that decays with increasing distance from the interface, the real parts of $k_z^{(d)}$ and $k_z^{(m)}$ must both be positive. Therefore, the only chance to equate the two sides of Eq. (1.13) is that $A = 0$ (and $B = 0$ as well). In other words, TE-polarized SPPs are not allowed to exist in the present system.

Before we proceed, it is well worth giving a further look at Eq. (1.4). Similar to how k_{sp} is related to the propagation length L_{sp}, the wavevectors perpendicular to the interface provide a measure of the decay length of the fields into each medium involved:

$$\delta_d = \frac{1}{\left|\mathrm{Im}\left\{k_z^{(d)}\right\}\right|}, \quad \delta_m = \frac{1}{\left|\mathrm{Im}\left\{k_z^{(m)}\right\}\right|}. \tag{1.14}$$

In the dielectric medium, δ_d is of the order of half the wavelength of the exciting light, while in the metal, δ_m is some tens of nanometers, comparable to the skin depth.

Knowing the decay lengths perpendicular to the interface helps to understand the SPPs on the surfaces of a metallic thin film (Fig. 1.3a), on the surfaces of a dielectric slab sandwiched between two metals, and in more sophisticated structures, such as alternating metal and dielectric multilayers. For simplicity, let us consider a metallic film with an air cladding and a dielectric substrate. The two interfaces between metal and air and between metal and the substrate each can support one SPP. When the thickness of the metallic film is much larger than the decay length δ_m, the SPPs at the two interfaces are decoupled and behave like those discussed above on the interface between a semi-infinite metal and a semi-infinite dielectric medium. For small thicknesses, the field confined at one of the interfaces can reach the other one and interacts with the field there before it decays to a negligible magnitude. This interaction between the fields at the two interfaces induces distortion of the dispersion curves. An example is shown in Fig. 1.3b for a 30-nm-thick silver film described by the free-electron model and on a glass substrate [4].

The dispersion relation of SPPs on a thin film can be derived by writing down the fields in each region as in the case of a single interface, but the solution is usually in an implicit form. One special solvable case is when the cladding and substrate are made of the same dielectric medium. The dispersion relation splits into two equations corresponding to two SPP modes that have odd and even parity in their longitudinal electric field (E_x) distributions, respectively. They are thus called odd and

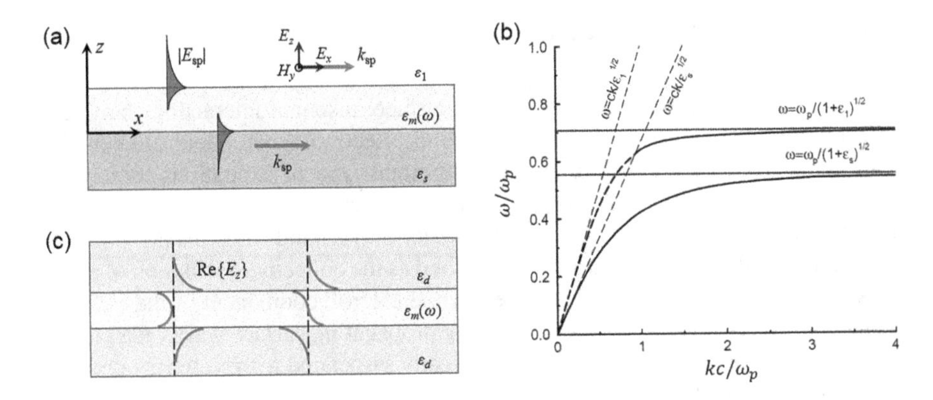

Fig. 1.3 (**a**) Propagation of SPPs on the surfaces of a thin metal film sandwiched between two dielectrics. (**b**) Dispersion diagram of the SPPs at the interfaces of a 30-nm-thick silver film with a vacuum cladding ($\varepsilon_1 = 1$) and a glass substrate ($\varepsilon_s = 2.25$). (**c**) Odd (left) and even (right) SPP modes on a metal film sandwiched by the same dielectric medium. Red curves denote the transverse electric field (E_z) profiles. (Panel (b) is reprinted with permission from [4]. Copyright (2005) Elsevier)

even modes. Meanwhile, the transverse fields (E_z and H_y) have the opposite parity, being even and odd for the two modes; see Fig. 1.3c. Again, when the metal is described by Eq. (1.7) and the wavenumber k is large, the asymptotic limits of the frequencies are as follows:

$$\omega_+ = \frac{\omega_p}{\sqrt{1+\varepsilon_d}}\sqrt{1+\frac{2\varepsilon_d \varepsilon^{-kd}}{1+\varepsilon_d}} \tag{1.15a}$$

for the odd mode, and

$$\omega_- = \frac{\omega_p}{\sqrt{1+\varepsilon_d}}\sqrt{1-\frac{2\varepsilon_d \varepsilon^{-kd}}{1+\varepsilon_d}}, \tag{1.15b}$$

for the even mode, with d the film thickness. Some papers base the naming on the symmetry of transverse fields, which leads to the terms of symmetric (for ω_+) and antisymmetric (for ω_-) modes. A less obvious but important property of film SPP modes is that the high-frequency, odd mode has lower attenuation as well as poorer field confinement than the SPPs at a single interface, whereas the low-frequency, even mode acts in the opposite way. Therefore, odd (even) modes can propagate over longer (shorter) distances and are also called long-range (short-range) SPPs [8]. Given the same metal and dielectric medium, the dispersion curve of the single interface SPP is sandwiched between the curves of the odd and even modes due to splitting. As k goes infinite in the absence of losses, the limiting frequencies converge to surface plasmon frequency ω_{sp}.

1.1.2 Localized Surface Plasmons

Besides SPPs at a planar metal-dielectric interface, resonant interactions between the oscillations of electrons in the metal and the electromagnetic field can exist in confined geometries, leading to another important type of excitations termed as localized surface plasmons (LSPs). Metallic nanoparticles that have a size much smaller than the wavelength are one of the major playgrounds of LSPs [9, 10]. The formation of LSPs can be intuitively understood as the collective oscillation of electrons in the metallic nanoparticles driven by the electric component of the electromagnetic field (Fig. 1.4a). Unlike SPPs being propagating surface waves at a planar interface, LSPs in confined geometries are mostly associated with scattering or radiating problems. In this section, we discuss the basic description and properties of LSPs.

We begin by considering a metallic sphere embedded in a dielectric medium, the simplest geometry that can be analytically solved. The dielectric functions of the metal and the surrounding medium are still $\varepsilon_m(\omega)$ and ε_d, respectively, as in the previous example. The interactions of this system with a plane wave can be obtained by using an exact electrodynamic approach, namely, the Mie theory [11–14]. However, we will keep it until the next section, where rigorous treatments are essential to unveil some unique properties of the dielectric spheres. When the sphere size a is much smaller than the wavelength of light λ, it is convenient to simplify and solve the problem under the electrostatic (or quasi-static) approximation [3, 4]. As its name stands, the electrostatic approximation treats the incident wave as an electrostatic field, which is uniform and has no phase variation (retardation) over the sphere. With this simplification, the problem can be described by the Laplace equation for a potential Φ:

$$\nabla^2\Phi = 0, \tag{1.16}$$

and the electric field is then determined by $\mathbf{E} = -\nabla\Phi$. In spherical coordinates, the general solution to Eq. (1.16) is

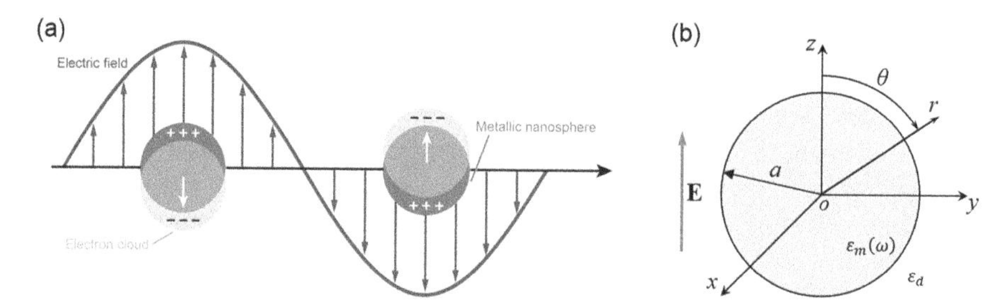

Fig. 1.4 (**a**) Oscillation of conduction electrons in a metallic nanosphere, driven by a time-harmonic electromagnetic field. (**b**) A metallic sphere with dielectric function $\varepsilon_m(\omega)$ embedded in an isotropic and homogeneous dielectric medium and placed in a uniform electrostatic field **E**

$$\Phi(r,\theta)=\sum_{l=0}^{\infty}\left[A_l r^l + B_l r^{-(l+1)}\right]P_l(\cos\theta), \tag{1.17}$$

if the system possesses azimuthal symmetry, as shown in Fig. 1.4b. The function $P_l(\cdot)$ denotes the Legendre polynomial of order l, and A_l and B_l are coefficients to be determined. The standard way of solving this boundary-value problem is to write down the potentials in both regions, i.e.,

$$\Phi^{(\text{in})}=\sum_{l=0}^{\infty}A_l^{(\text{in})}r^l P_l(\cos\theta) \tag{1.18a}$$

for inside the sphere ($r < a$) due to the finiteness of the potential at the origin,

$$\Phi^{(\text{out})}=\sum_{l=0}^{\infty}\left[A_l^{(\text{out})}r^l + B_l^{(\text{out})}r^{-(l+1)}\right]P_l(\cos\theta), \tag{1.18b}$$

for outside the sphere ($r \geq a$), and impose proper boundary conditions at $r = a$ and $r = \infty$. For an electrostatic field $\mathbf{E} = E_0\hat{\mathbf{z}}$, it can be proven that the potentials are

$$\Phi^{(\text{in})}=-\frac{3\varepsilon_d}{\varepsilon_m+2\varepsilon_d}E_0 r\cos\theta, \tag{1.19a}$$

$$\Phi^{(\text{out})}=-E_0 r\cos\theta+\frac{\varepsilon_m-\varepsilon_d}{\varepsilon_m+2\varepsilon_d}E_0 a^3\frac{\cos\theta}{r^2}. \tag{1.19b}$$

Equation (1.19b) is composed of two terms. The first term is nothing else but the potential of the applied field, and the second term can be related to an induced electric dipole at the center of the sphere, which is maximized when the modulus of the denominator $|\varepsilon_m + 2\varepsilon_d|$ is minimal, bringing in the resonant characteristic. This dipole resonance is the LSPs of the lowest order. When the loss of the metal is not significant, a more practical expression of the resonance condition is

$$\text{Re}\{\varepsilon_m(\omega)\}=-2\varepsilon_d, \tag{1.20}$$

while for a metal described by the free-electron model in Eq. (1.7) with no loss, the resonance simply diverges at the frequency $\omega_1=\omega_p/\sqrt{1+2\varepsilon_d}$.

An alternative approach for the analysis of LSPs is to consider the eigenmodes of the structure without applying the external field [15]. Under the electrostatic approximation, the solutions to Eq. (1.16) for the l-th order mode are

$$\Phi_l^{(\text{in})}=A\left(\frac{r}{a}\right)^l P_l(\cos\theta), \tag{1.21a}$$

$$\Phi_l^{(\text{out})} = A\left(\frac{r}{a}\right)^{-(l+1)} P_l(\cos\theta), \tag{1.21b}$$

where A is a normalization constant. The solutions are constituted to naturally fulfill the continuity requirement on the tangential electric fields at $r = a$, and by requiring the normal components of the electric displacement $-\varepsilon(\partial\Phi_l/\partial r)$ to be equal, we get

$$l\varepsilon_m(\omega) + (l+1)\varepsilon_d = 0. \tag{1.22}$$

Inserting Eq. (1.7) into Eq. (1.22), the frequency of the l-th order mode (or multipole) is given by

$$\omega_l = \omega_p\sqrt{\frac{l}{l+(l+1)\varepsilon_d}}. \tag{1.23}$$

It can now be seen that the resonance condition ω_1 after Eq. (1.20) is the mode frequency for the dipole that has $l = 1$. Notice the mode frequencies are functions of the permittivity of the dielectric medium. The spectral shifts in response to the refractive index change in the environment are the key information for plasmonic sensing. Quasi-static resonance frequencies are independent of the sphere size. Practically, higher-order modes ($l > 1$) arise as the particle size increases or at higher frequencies, showing multiple nodes in the surface charge density distribution, as illustrated in Fig. 1.5. For very large mode indices ($l \to \infty$), the mode frequencies in Eq. (1.23) approach the surface plasmon frequency ω_{sp} in Eq. (1.9).

Next, because the dipole mode has several important properties related to the later discussion on optical antennas, we brief the reader on the concepts of near field and far field, which have frequent appearance in the context of nanophotonics and can be well defined on dipoles. In a dielectric medium of permittivity ε_d, the electric field of an electric dipole at the origin can be expressed by [17].

$$\mathbf{E}(\mathbf{r}) = \frac{1}{4\pi\varepsilon_0\varepsilon_d}\frac{e^{ikr}}{r}\left\{k^2\left[(\mathbf{n}\times\mathbf{p})\times\mathbf{n}\right] + \left[3\mathbf{n}(\mathbf{n}\cdot\mathbf{p}) - \mathbf{p}\right]\frac{1}{r}\left(\frac{1}{r} - ik\right)\right\}e^{-i\omega t} \tag{1.24}$$

in the spherical coordinate system. Here, $\mathbf{p}$ is the electric dipole moment, k is the wavenumber in the dielectric medium, and $\mathbf{n}$ is the unit vector in the direction of $\mathbf{r}$. The near field is regarded as the zone where $kr \ll 1$. The second term in the braces, as a result, dominates the field. It can be shown that the electrostatic field associated with the second term in Eq. (1.19b) has the expression

$$\mathbf{E}(\mathbf{r}) = \frac{1}{4\pi\varepsilon_0\varepsilon_d}\frac{3\mathbf{n}(\mathbf{n}\cdot\mathbf{p}) - \mathbf{p}}{r^3}, \tag{1.25}$$

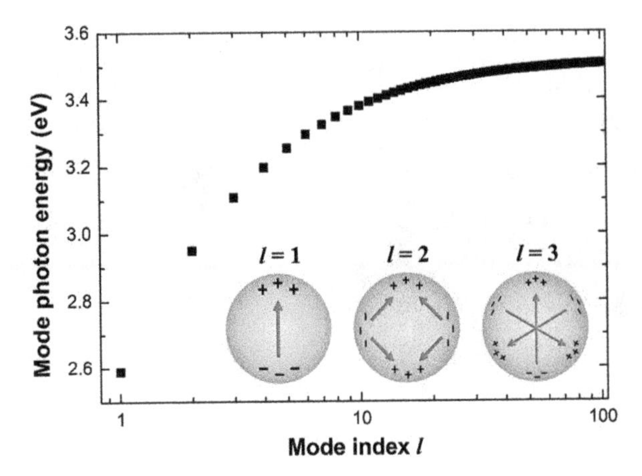

Fig. 1.5 Frequencies of LSPs versus mode index l for a gold nanosphere embedded in a dielectric medium with $n = 2.35$. The dielectric function of gold is described by the free-electron model in Eq. (1.7) with the plasma frequency taken from [16]. Insets sketch the charge distributions on the gold surface for the first three modes $l = 1, 2, 3$. Net dipole moment (blue arrows) exists only for $l = 1$

with

$$\mathbf{p} = 4\pi\varepsilon_0\varepsilon_d a^3 \frac{\varepsilon_m - \varepsilon_d}{\varepsilon_m + 2\varepsilon_d}\mathbf{E}_0 = \varepsilon_0\varepsilon_d \alpha\,\mathbf{E}_0. \tag{1.26}$$

The second equator in Eq. (1.26) defines the polarizability α of the sphere under the quasi-static approximation. By comparing these equations with Eq. (1.24) in the limit of $kr \to 0$, we can conclude that apart from the harmonic oscillation in time, the fields within the near-field region of an electric dipole (or dipole mode) have the properties of electrostatic fields. Figure 1.6a illustrates the electric field intensity around a plasmonic nanosphere at dipole resonance, where the fields are tightly localized near the sphere, with the distribution elongated along the dipole moment **p**.

The far field, on the other hand, is defined by the opposite limit $kr \gg 1$. In this region, the electric field is dominated by the first term in the braces of Eq. (1.24), which is transverse to the radial direction, showing the typical behavior of radiation fields. Therefore, the far field is also referred to as the radiation zone. Within the far field, the radiation pattern and field distribution of a resonant plasmonic nanosphere resemble those of an electric dipole, which, in contrast to the near field, extend in the directions perpendicular to the dipole moment **p**, as shown in Fig. 1.6b. The properties of LSPs set the basis for the use of resonant nanoparticles in optical antennas and metasurfaces as building blocks.

With all the preparations above based on spheres, we close this section with a quick glance at LSPs in other geometries. Although the mode frequencies under the quasi-static approximation are independent of the particle size, in practical circumstances, the approximation is not strictly valid, and we will see in the next section

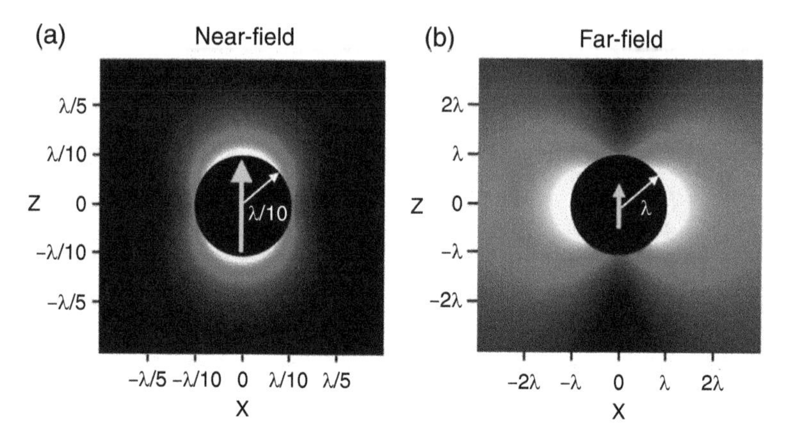

Fig. 1.6 (**a**) Electric field intensity around a resonant plasmonic nanosphere. (**b**) Electric field intensity at larger distances from a resonant plasmonic nanosphere. The black disks denote fictious spheres enclosing the particles, and the green arrows indicate the positions of the induced electric dipole. (Reprinted from [18] with permission of Springer Nature)

how the size parameter comes into play. In addition to size and materials, the shape of the particle will also influence the resonances. For particles elongated in certain directions, e.g., spheroids and rods, the mode frequencies will have dependence on the polarization direction of the incident field with respect to each principal axis. This can be seen from Fig. 1.7. When a rod is aligned parallel or perpendicular to the polarization direction, the incident electric field drives the electrons in the rod to oscillate over different lengths, resulting in two distinct resonances. The mode frequencies and polarizabilities are determined by formulas similar to Eqs. (1.20) and (1.26), where the shape information is encoded via some geometric depolarization factors in the polarizability of the particle [12].

The quasi-static approximation fails to be valid when a nanoparticle has at least one dimension close to or greater than the wavelength of light. In this case, because the field across the particle is no longer uniform, the retardation effect needs to be considered [20]. Three examples of field intensity distribution around silver strips involving different dimensions are compared in Fig. 1.8. At normal incidence, the strip of subwavelength width acts just like a dipole in the cross-sectional plane. For wavelength-scale strips, the patterns show multiple nodes as the high-order modes do, but they are formed through the constructive interference of short-range SPPs bouncing back and forth between the edges, like the Fabry-Pérot resonances.

1.1.3 Excitation of Surface Plasmon Polaritons by Light

Lastly, we briefly discuss the excitation of SPPs in experiments. Opposed to LSPs that can be excited by simple light illumination, launching SPPs on a metal-dielectric interface needs special techniques. As discussed above, the dispersion curve of an

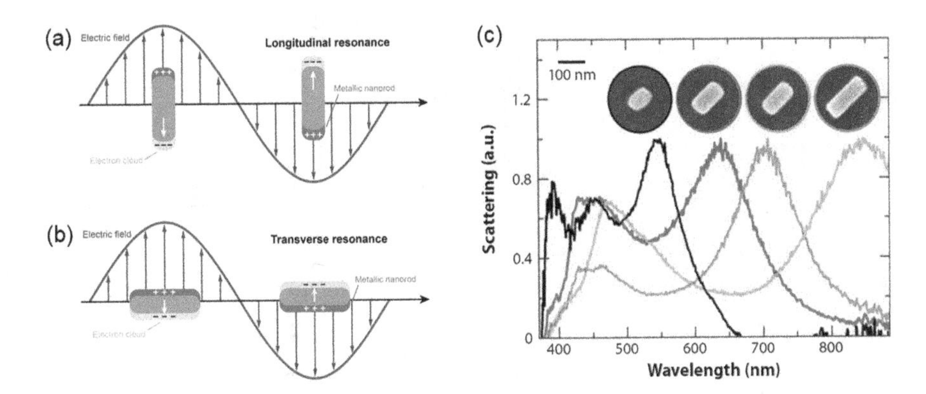

Fig. 1.7 LSPs in nanorods. Oscillation of conduction electrons in a metallic nanorod driven by an electromagnetic field polarized along (**a**) and perpendicular to (**b**) the long axis leads to longitudinal and transverse resonances, respectively. (**c**) Measured scattering spectra of individual silver nanobars. For each spectrum, the peak at a longer (shorter) wavelength corresponds to the longitudinal (transverse) plasmon resonance, except a substrate-induced splitting of the transverse peak for the nanocube (black curve). (Panel (c) is reprinted with permission from [19]. Copyright (2007) American Chemical Society)

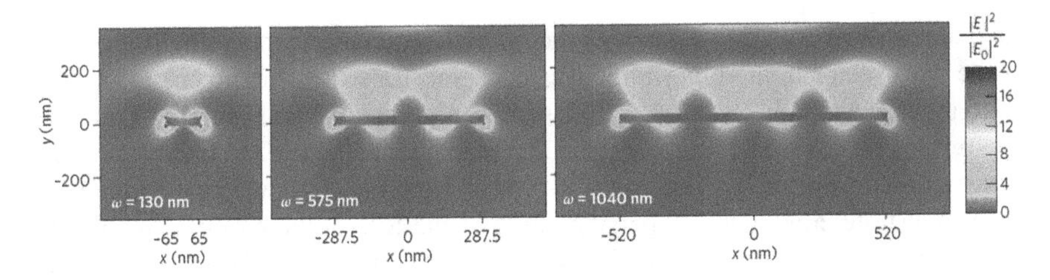

Fig. 1.8 Electric field intensity distributions for the three lowest-order resonances $l = 1, 3, 5$ in 30-nm-thick silver strips. The incident field has a wavelength $\lambda = 550$ nm and is polarized along the x-axis. At normal incidence, only odd order resonances can be excited because of symmetry. (Reprinted from [20, 21] with permissions of Springer Nature and Optica Publishing Group (© 2018))

SPP lies to the right of the light line of the constituting dielectric medium. In other words, the wavenumber of the SPP is always larger than the wavenumber of the exciting light, meaning that an extra momentum along the interface must be provided to couple the incident light to an SPP. Despite a variety of techniques achieving this [22], the following discussion is limited to the subcategory of optical means [4].

One of the most widely used strategies is coupling via a prism based on attenuated total reflection (ATR). This is particularly useful for thin metal films. A representative geometry known as the Kretschmann configuration [23] is illustrated in Fig. 1.9a, where a dielectric prism is attached to the bottom of a thin metal film, forming the structure we have seen in Fig. 1.3. Although SPPs cannot be excited by simple illumination from the medium constituting the interface because of the

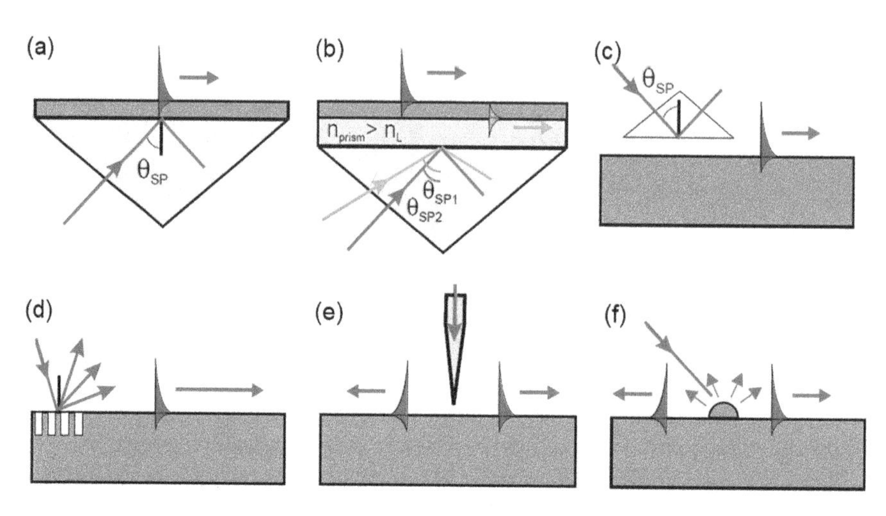

Fig. 1.9 Optical excitation of SPPs. (**a–c**) Prism coupling in the Kretschmann configuration (**a, b**) and Otto configuration (**c**). (**d**) Grating coupling. (**e, f**) Near-field excitation by diffraction from an aperture (**e**) and small scatterers on the surface (**f**). (Reprinted with permission from [4]. Copyright (2005) Elsevier)

mismatch of wavevectors, in an asymmetric dielectric environment, the projection of the wavevector in the high-refractive-index substrate (i.e., the prism) along the bottom surface $k_x = k_{prism} \cdot \sin\theta = n_{prism} \cdot k_0 \cdot \sin\theta$ can exceed the wavenumber k_0 in the low-refractive-index cladding (e.g., air) and match k_{sp} at the top interface for a certain angle of incidence θ_{sp}. This can also be seen from Fig. 1.3b, where the light line of the substrate crosses the dispersion curve of the SPP at the air/metal interface. The excitation of SPP thus happens as a result of the tunneling of evanescent waves from ATR through the metal film when $\theta \approx \theta_{sp}$; otherwise the incident light is reflected off the metal/prism interface. In experiments, the excitation of SPP is identified as a minimum in the reflection intensity. The Kretschmann configuration can be varied by introducing another thin layer of dielectric medium with $n_1 < n_{prism}$ between the metal and the prism (Fig. 1.9b). This adaption allows excitation of SPPs at both interfaces of the metal film if its thickness is smaller than the decay lengths. In occasions where the metal is thick or direct contact with a prism is undesired, SPPs can be accessed by using the Otto configuration [24]. As illustrated in Fig. 1.9c, the prism is separated from the metal film by a small air gap, across which the evanescent wave from ATR at the prism/air interface is coupled to SPPs at the air/metal interface.

Another method, which is straightforward and can be simply formulated, is coupling by a grating. The schematic is illustrated in Fig. 1.9d. According to the theory of optical diffraction, a one-dimensional (1D) periodic array of grooves with lattice constant a can provide wavevectors along the direction of periodicity $\boldsymbol{x}$, with the amplitude being integer multiples of the reciprocal lattice vector $\mathbf{G} = (2\pi/a)\ \boldsymbol{x}$. Therefore, at the angle of incidence θ, the projection of the wavevector of the incident light along $\boldsymbol{x}$ is offset by the extra wavevectors from the grating [25]:

$$k_x = k_0 \sin\theta + j\frac{2\pi}{a}, \tag{1.27}$$

where j is an integer, and SPPs can be excited when k_x equals k_{sp}.

Diffraction from other structures, for example, an aperture at the tip of a probe or the end of a fiber (Fig. 1.9e) and scatterers on the metal surface (Fig. 1.6f), can also be used to excite SPPs. These structures own the advantage of local excitation as they are subwavelength in size, and the reverse process allows high-resolution imaging of surface features, such as near-field scanning optical microscopy (NSOM). However, because the energy of the incident light is distributed over all the diffracted components with a broad range of wavevectors, the excitation efficiency is usually lower than that of prism or grating coupling.

1.2 Mie Resonances in Dielectric Nanostructures

Plasmonics have enabled unprecedented opportunities in nanophotonics. However, the associated losses in metals, which convert a considerable amount of optical energy into heat through light absorption, could impede many applications. Resonant nanostructures made of high-refractive-index dielectric materials and semiconductors featuring low losses offer a different approach to manipulating light at the nanoscale [26]. In this section, we introduce the fundamentals of Mie theory for the analysis of dielectric resonators.

1.2.1 Mie Theory

The Mie theory, named after one of the pioneers who developed its formalism [27], gives the exact solution to the scattering of electromagnetic waves by homogeneous spheres. The real power of Mie theory is nevertheless far beyond this. Remarkable insights into scattering by small particles become accessible in light of Mie theory. Therefore, in some subfields of nanophotonics such as nanoantennas and metasurfaces, it has been extensively used in the study of many diverse types of scatterers and resonators. This section is devoted to the mathematical basis of the Mie theory. For the history, development, theory in detail, and applications, we direct the interested reader to the classic works by Kerker [11] and by Bohren and Huffman [12], among many other excellent collections [28].

The standard problem of Mie scattering is illustrated in Fig. 1.10. A sphere of radius a is embedded in a linear, homogeneous, and isotropic host medium of refractive index $n_h = \sqrt{\varepsilon_h \mu_h}$, with ε_h and μ_h being the relative permittivity and permeability. Similarly, the refractive index of the sphere is $n_s = \sqrt{\varepsilon_s \mu_s}$. The incident plane

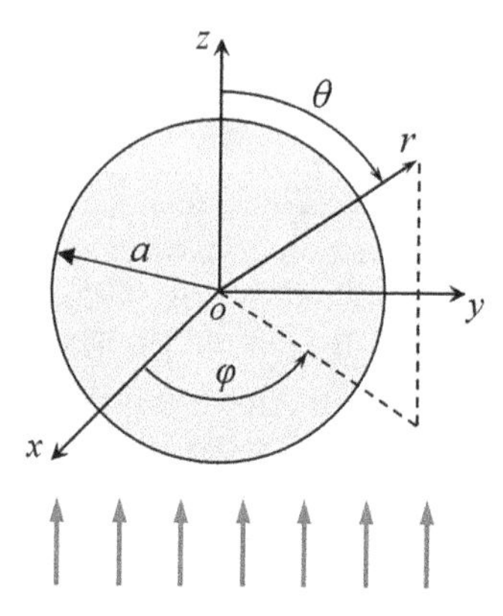

Fig. 1.10. A plane wave propagating in the +*z* direction illuminates on a sphere of radius *a*, centered at the origin of the spherical coordinate system

wave is monochromatic, time-harmonic, and arbitrarily polarized, and its electric and magnetic field components are

$$\mathbf{E}_i = \mathbf{E}_0 \exp\left(i\mathbf{k}_h \cdot \mathbf{r} - i\omega t\right), \quad \mathbf{H}_i = \mathbf{H}_0 \exp\left(i\mathbf{k}_h \cdot \mathbf{r} - i\omega t\right), \tag{1.28}$$

where $\mathbf{k}_h$ is the wavevector in the host medium, $\boldsymbol{r}$ is position vector in space, and ω is the angular frequency of the fields. We denote the field inside the sphere by ($\mathbf{E}_1$, $\mathbf{H}_1$) and the field outside the sphere in the surrounding medium by ($\mathbf{E}_2$, $\mathbf{H}_2$). When solving the scattering problem, a common strategy is to write the total field in the unbounded region into two parts: the incident field ($\mathbf{E}_i$, $\mathbf{H}_i$) and scattered field ($\mathbf{E}_s$, $\mathbf{H}_s$). In the present case, these fields are connected through

$$\mathbf{E}_2 = \mathbf{E}_i + \mathbf{E}_s, \quad \mathbf{H}_2 = \mathbf{H}_i + \mathbf{H}_s. \tag{1.29}$$

The fields mentioned above should satisfy the vector wave equations

$$\nabla^2 \mathbf{E} + k^2 \mathbf{E} = 0, \quad \nabla^2 \mathbf{H} + k^2 \mathbf{H} = 0, \tag{1.30}$$

at all points inside and outside the sphere, with k the wavenumber in the corresponding medium. All the subscripts are omitted. The general solution to Eq. (1.30), i.e., the vector spherical harmonics $\boldsymbol{M}$ and $\boldsymbol{N}$, can be accomplished with some neat mathematical treatment and expressed by

$$\mathbf{M}_{\mathrm{emn}}^{(l)} = -\frac{m}{\sin\theta} \sin m\varphi P_n^m\left(\cos\theta\right) z_n\left(kr\right) \hat{\mathbf{e}}_\theta - \cos m\varphi \frac{dP_n^m\left(\cos\theta\right)}{d\theta} z_n\left(kr\right) \hat{\mathbf{e}}_\varphi, \tag{1.31a}$$

$$\mathbf{M}_{\mathrm{omn}}^{(l)} = \frac{m}{\sin\theta} \cos m\varphi P_n^m\left(\cos\theta\right) z_n\left(kr\right) \hat{\mathbf{e}}_\theta - \sin m\varphi \frac{dP_n^m\left(\cos\theta\right)}{d\theta} z_n\left(kr\right) \hat{\mathbf{e}}_\varphi, \tag{1.31b}$$

$$\mathbf{N}_{\text{emn}}^{(l)} = n(n+1)\cos m\varphi P_n^m(\cos\theta)\frac{z_n(kr)}{kr}\hat{\mathbf{e}}_r + \cos m\varphi \frac{dP_n^m(\cos\theta)}{d\theta}\frac{1}{kr}\frac{d}{dr}[rz_n(kr)]\hat{\mathbf{e}}_\theta - m\sin m\varphi \frac{P_n^m(\cos\theta)}{\sin\theta}\frac{1}{kr}\frac{d}{dr}[rz_n(kr)]\hat{\mathbf{e}}_\varphi, \tag{1.31c}$$

$$\mathbf{N}_{\text{omn}}^{(l)} = n(n+1)\sin m\varphi P_n^m(\cos\theta)\frac{z_n(kr)}{kr}\hat{\mathbf{e}}_r + \sin m\varphi \frac{dP_n^m(\cos\theta)}{d\theta}\frac{1}{kr}\frac{d}{dr}[rz_n(kr)]\hat{\mathbf{e}}_\theta + m\cos m\varphi \frac{P_n^m(\cos\theta)}{\sin\theta}\frac{1}{kr}\frac{d}{dr}[rz_n(kr)]\hat{\mathbf{e}}_\varphi. \tag{1.31d}$$

Here, P_n^m is the associated Legendre polynomial of the nth degree and mth order; z_n is one of the four spherical Bessel and Hankel functions j_n, y_n, $h_n^{(1)} = j_n + iy_n$, and $h_n^{(2)} = j_n - iy_n$; and $\hat{e}_s$ ($s = r, \theta, \varphi$) is the unit vector in the radial (r), polar (θ), or azimuthal (φ) direction. Two indices of $\boldsymbol{M}$ and $\boldsymbol{N}$ that do not explicitly appear on the right side are l in the superscript, which takes its value from 1 to 4 to indicate the choice of z_n from the four options above in the given order, and e/o in the subscript, which denotes the even/odd dependence of the spherical harmonics on the azimuthal angle φ.

After some lengthy algebra, the incident field can be expressed by a series of vector spherical harmonics as

$$\mathbf{E}_i = E_0 \sum_{n=1}^{\infty} i^n \frac{2n+1}{n(n+1)}\left(\mathbf{M}_{\text{o1n}}^{(1)} - i\mathbf{N}_{\text{e1n}}^{(1)}\right), \tag{1.32a}$$

$$\mathbf{H}_i = -\frac{k_h}{\omega\mu_0\mu_h} E_0 \sum_{n=1}^{\infty} i^n \frac{2n+1}{n(n+1)}\left(\mathbf{M}_{\text{e1n}}^{(1)} + i\mathbf{N}_{\text{o1n}}^{(1)}\right). \tag{1.32b}$$

By aligning the center of the sphere to the origin of the spherical coordinates, the expansions of the fields inside the particle are

$$\mathbf{E}_1 = E_0 \sum_{n=1}^{\infty} i^n \frac{2n+1}{n(n+1)}\left(c_n\mathbf{M}_{\text{o1n}}^{(1)} - id_n\mathbf{N}_{\text{e1n}}^{(1)}\right), \tag{1.33a}$$

$$\mathbf{H}_1 = -\frac{k_s}{\omega\mu_0\mu_s} E_0 \sum_{n=1}^{\infty} i^n \frac{2n+1}{n(n+1)}\left(d_n\mathbf{M}_{\text{e1n}}^{(1)} + ic_n\mathbf{N}_{\text{o1n}}^{(1)}\right) \tag{1.33b}$$

with $k_s = (n_s/n_h)\cdot k_h$, and the scattered fields are

$$\mathbf{E}_s = E_0 \sum_{n=1}^{\infty} i^n \frac{2n+1}{n(n+1)} \left(ia_n \mathbf{N}_{\mathrm{e1n}}^{(3)} - b_n \mathbf{M}_{\mathrm{o1n}}^{(3)} \right), \tag{1.34a}$$

$$\mathbf{H}_s = -\frac{k_h}{\omega \mu_0 \mu_h} E_0 \sum_{n=1}^{\infty} i^n \frac{2n+1}{n(n+1)} \left(ib_n \mathbf{N}_{\mathrm{o1n}}^{(3)} + a_n \mathbf{M}_{\mathrm{e1n}}^{(3)} \right). \tag{1.34b}$$

The choice of j_n in Eqs. (1.32) and (1.33) is required by the finiteness of values at the origin, while the necessity of having $h_n^{(1)}$ in the scattered fields is less obvious. With the asymptotic expressions of the spherical Bessel and Hankel functions for large values of $k_h \cdot r$, one will see that only $h_n^{(1)}$ corresponds to an outgoing spherical wave in accordance with the physical meaning of "scattering."

The undefined variables a_n through d_n in Eqs. (1.33) and (1.34) are the Mie coefficients. Their expressions can be derived from the boundary conditions at the surface of the sphere, which require the tangential components of ($\mathbf{E}_1$, $\mathbf{H}_1$) and ($\mathbf{E}_2$, $\mathbf{H}_2$) to be continuous. Here, we only show the two coefficients associated with the scattered fields:

$$a_n = \frac{\mu_h m^2 j_n(mx)\left[xj_n(x)\right]' - \mu_s j_n(x)\left[mxj_n(mx)\right]'}{\mu_h m^2 j_n(mx)\left[xh_n^{(1)}(x)\right]' - \mu_s h_n^{(1)}(x)\left[mxj_n(mx)\right]'}, \tag{1.35a}$$

$$b_n = \frac{\mu_s j_n(mx)\left[xj_n(x)\right]' - \mu_h j_n(x)\left[mxj_n(mx)\right]'}{\mu_s j_n(mx)\left[xh_n^{(1)}(x)\right]' - \mu_h h_n^{(1)}(x)\left[mxj_n(mx)\right]'}, \tag{1.35b}$$

where $x = k_h \cdot a$ is the size parameter, $m = n_s/n_h$ is relative refractive index, and the prime denotes a derivative. If the sphere and the surrounding medium are both nonmagnetic, $\mu_s = \mu_h = 1$, and the coefficients can be reformulated to

$$a_n = \frac{m\varphi_n(mx)\varphi_n'(x) - \varphi_n(x)\varphi_n'(mx)}{m\varphi_n(mx)\zeta_n'(x) - \zeta_n(x)\varphi_n'(mx)}, \tag{1.36a}$$

$$b_n = \frac{\varphi_n(mx)\varphi_n'(x) - m\varphi_n(x)\varphi_n'(mx)}{\varphi_n(mx)\zeta_n'(x) - m\zeta_n(x)\varphi_n'(mx)}, \tag{1.36b}$$

with j_n and $h_n^{(1)}$ replaced by the corresponding Riccati-Bessel functions φ_n and ζ_n, respectively.

The Mie coefficients are not defined just for the convenience of expression but have physical meanings. Recalling Eq. (1.34), the terms of electric field containing a_n have a radial component from $\mathbf{N}_{e1n}^{(3)}$, whereas the magnetic field components with a_n are purely transverse to the radial direction. Therefore, the coefficients a_n represent the transverse magnetic (TM) components of the scattered waves. This leads to a more intuitive interpretation of the coefficients a_n, in which they are associated with a class of field structures or "modes," known as the electric multipoles. When $n = 1, 2, \ldots$, the modes are in turn electric dipole (ED), electric quadrupole (EQ), and subsequent higher-order ones. In the same way, the coefficients b_n are connected to the transverse electric (TE) components of scattering and the magnetic multipoles. The field line patterns for the first four multipoles of a dielectric sphere are shown in Fig. 1.11.

The presence of a particle in an incident wave will disturb the original power flow, resulting in extinction of the incident beam. If the host medium is nonabsorbing, the extinction can be broken down into two parts: absorption and scattering, both by the particle. This can be seen more clearly with the following analysis. Outside the particle, the time-averaged Poynting vector of the total field is

$$\mathbf{S} = \frac{1}{2}\mathrm{Re}\left\{\mathbf{E}_2 \times \mathbf{H}_2^*\right\} = \frac{1}{2}\mathrm{Re}\left\{\left(\mathbf{E}_i + \mathbf{E}_s\right)\times\left(\mathbf{H}_i^* + \mathbf{H}_s^*\right)\right\}. \tag{1.37}$$

The right side of the equation can be decomposed to three terms:

$$\begin{aligned} \mathbf{S}_i &= \frac{1}{2}\mathrm{Re}\left\{\mathbf{E}_i \times \mathbf{H}_i^*\right\}, \quad \mathbf{S}_s = \frac{1}{2}\mathrm{Re}\left\{\mathbf{E}_s \times \mathbf{H}_s^*\right\}, \\ \mathbf{S}_{\text{ext}} &= \frac{1}{2}\mathrm{Re}\left\{\mathbf{E}_i \times \mathbf{H}_s^* + \mathbf{E}_s \times \mathbf{H}_i^*\right\}, \end{aligned} \tag{1.38}$$

corresponding to the components from the incident field, the scattered field, and their interference, respectively. Consider the particle is enclosed by an imaginary surface A. From the conservation law, the rate W_a at which energy is absorbed by the particle is equal to the net rate of energy flowing across the surface A:

$$W_a = -\int_A \mathbf{S}\cdot\hat{e}_r dA = -W_i - W_s + W_{\text{ext}}, \tag{1.39}$$

where

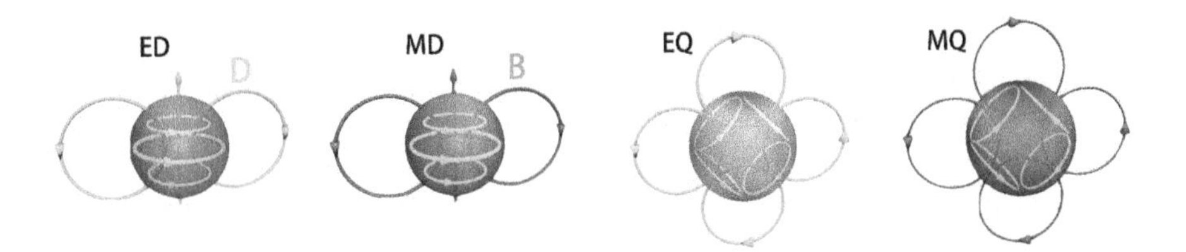

Fig. 1.11 Field structures of four multipoles of the lowest order supported by a dielectric sphere, namely, electric dipole (ED) from a_1, magnetic dipole (MD) from b_1, electric quadrupole (EQ) from a_2, and magnetic quadrupole (MQ) from b_2. Curves in orange denote the electric displacement lines, and curves in light blue represent the magnetic field lines. (Reprinted from [26] with permission from AAAS)

$$W_i = \int_A \mathbf{S}_i \cdot \hat{e}_r dA, \quad W_s = \int_A \mathbf{S}_s \cdot \hat{e}_r dA, \quad W_{\text{ext}} = -\int_A \mathbf{S}_{\text{ext}} \cdot \hat{e}_r dA. \tag{1.40}$$

Because W_i is zero in non-absorbing media, Eq. (1.39) can be rewritten as

$$W_{\text{ext}} = W_s + W_a, \tag{1.41}$$

an expression that shows extinction is the sum of scattering and absorption. Furthermore, if each term in this equation is divided by the incident irradiance I_0, we get, without walking through the derivation, the corresponding cross sections with the dimension of area. Specifically, the scattering cross section of a sphere is

$$C_{\text{sct}} = \frac{W_s}{I_0} = \frac{2\pi}{k_0^2} \sum_{n=1}^{\infty} (2n+1)\left(|a_n|^2 + |b_n|^2\right), \tag{1.42}$$

the extinction cross section is

$$C_{\text{ext}} = \frac{W_{ext}}{I_0} = \frac{2\pi}{k_0^2} \sum_{n=1}^{\infty} (2n+1)\text{Re}\{a_n + b_n\}, \tag{1.43}$$

and the absorption cross section is just their difference. It is also convenient to define efficiencies by further dividing cross sections by the cross-sectional area of the sphere, resulting in

$$Q_{\text{sct}} = \frac{C_{\text{sct}}}{\pi a^2} = \frac{2}{x^2} \sum_{n=1}^{\infty} (2n+1)\left(|a_n|^2 + |b_n|^2\right), \tag{1.44}$$

$$Q_{\text{ext}} = \frac{C_{\text{ext}}}{\pi a^2} = \frac{2}{x^2} \sum_{n=1}^{\infty} (2n+1)\text{Re}\{a_n + b_n\}. \tag{1.45}$$

Another useful quantity describing the efficiency for backscattering is given by [12].

$$Q_b = \frac{1}{x^2} \left| \sum_{n=1}^{\infty} (2n+1)(-1)^n (a_n - b_n) \right|^2. \tag{1.46}$$

These dimensionless quantities provide a measure of the scattering, absorption, and extinction capability of a particle with respect to its physical size. This allows us to go beyond the picture of geometric optics to understand the interactions between a particle and the wave irradiating on it. Figure 1.12 compares the scattering efficiencies of silicon spheres in different contexts. In Fig. 1.12a, a 200-nm crystalline sphere is characterized in the visible region. It is evident that the efficiency curve is well above unity over the entire range, especially around certain wavelengths that exhibit sharp spectral features. By evaluating the contributions by each term in Eq. (1.44), these features can be related to the Mie coefficients one by one: when the denominator of a Mie coefficient is close to zero, the term

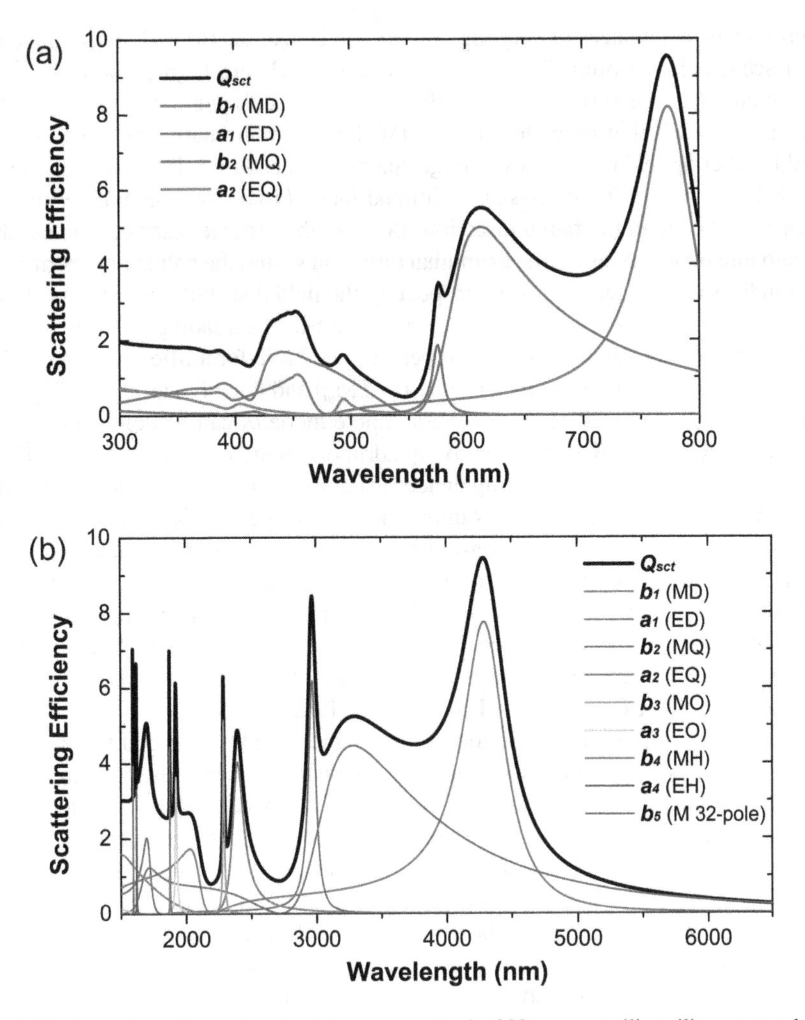

Fig. 1.12 Scattering efficiencies (thick black curves) of a 200-nm crystalline silicon nanosphere (**a**) and a 1200-nm silicon microsphere (**b**). Contributions by individual Mie resonances are plotted in thinner curves. Only four resonances can be effectively excited in the first case due to loss, whereas modes up to 32-pole are recognized in the second case, where the relative permittivity of silicon is taken as a constant $\varepsilon_{\mathrm{Si}} = 11.7$. EO/MO, electric/magnetic octupole. EH/MH, electric/magnetic hexadecapole

overpowers others in the expansion, and the corresponding mode will dominate the scattering in a resonant behavior. Therefore, the multipolar modes we discussed above also appear frequently as "Mie resonances" in the literature. In the visible region, silicon has non-negligible loss [29]. This hinders the occurrence of high-order modes at short wavelengths. However, for a particle made of lossless high-refractive-index material, such as a micrometer-sized silicon sphere in the infrared region, high-order modes manifest themselves as sharp peaks with decreasing line-widths, as shown in Fig. 1.12b. In even larger spheres sized some tens of micrometers, resonances of very high orders can be excited through phase-matched evanescent field coupling [30]. In the last case, fields are highly confined in the sphere and near the surface, self-interfering while circulating along the perimeter, known as the whispering gallery modes (WGMs). An alternative notation is often used to label optical resonances in large spheres. Instead of only using n, the order of a Mie resonance, it employs an additional index l to denote the number of field intensity maxima in the radial direction. Because the order n describes the number of field intensity maxima in the azimuthal direction within the half sphere perimeter, both indices can be determined by inspecting the field distributions [31]. As shown in Fig. 1.12, every Mie coefficient has multiple peaks spreading towards the left, and the index l corresponds to the l-th peak of the curve for a Mie coefficient. This allows each peak of a_n to be labeled as a_{nl} (or TM_{nl}) and that of b_n as b_{nl} (or TE_{nl}). For example, the major peak of the magnetic dipole mode b_1 can be denoted by b_{11} or TE_{11}, and the first peak of the electric quadrupole mode a_2 becomes a_{21} or TM_{21}. Most nanophotonic study involving dielectric components deals with low-order Mie resonances with $n \leq 2$ and $l = 1$. Naming them with Mie coefficients or multipoles is thus clear enough and more convenient. Nonetheless, the two-index labeling is favorable when a dielectric resonator is optically large and supports spectrally close high-order resonances [32], such as WGMs. For plasmonic nanoparticles that have a negative permittivity, the coefficients b_n are usually vanishingly small. Consequently, the optical responses are dominated by electric multipoles, as what we have seen in the discussion of LSPs in Sect. 1.1.2.

The above framework of Mie theory for spheres can be extended to concentric multilayer spheres and nonspherical particles such as spheroids, as long as the boundaries coincide with the surfaces of a coordinate system where the wave equation is solvable by separation of variables. For particles with irregular shape, scattering problems are usually dealt with by using numerical methods. In this case, a very useful tool for gaining physical insights is the multipole expansion, which extracts multipole moments of the resonant particle from numerically computed field distributions. Despite several general forms of exact expressions in classic textbooks [17], here we introduce an alternative faormulation more manageable in the study of nanophotonics [33, 34]. For a given scattering problem in a vacuum, once the total electric field $\mathbf{E}$ is known, the induced electric current density $\mathbf{J}$ in the scatterer (i.e., the particle) is given by $\mathbf{J} = -i\omega\varepsilon_0(\varepsilon_r - 1)\mathbf{E}$. This current density $\mathbf{J}$ is the source that generates all the multipole fields outside the particle. Subsequently, the individual multipole moments can be expressed in terms of proper integrals over $\mathbf{J}$. The results of the four lowest orders of multipoles, namely, electric dipole $\mathbf{p}$, magnetic dipole $\mathbf{m}$, electric quadrupole $\mathbf{Q}^{\mathrm{e}}$, and magnetic quadrupole $\mathbf{Q}^{\mathrm{m}}$, are

$$p_\alpha = -\frac{1}{i\omega}\left\{\int d^3\mathbf{r} J_\alpha j_0(k_0 r) + \frac{k_0^2}{2}\int d^3\mathbf{r}\left[3(\mathbf{r}\cdot\mathbf{J})r_\alpha - r^2 J_\alpha\right]\frac{j_2(k_0 r)}{(k_0 r)^2}\right\}, \tag{1.47a}$$

$$m_\alpha = \frac{3}{2}\int d^3\mathbf{r}(\mathbf{r}\times\mathbf{J})_\alpha \frac{j_1(k_0 r)}{k_0 r}, \tag{1.47b}$$

$$Q^e_{\alpha\beta} = -\frac{3}{i\omega}\left\{\begin{array}{l} \int d^3\mathbf{r}\left[3\left(r_\beta J_\alpha + r_\alpha J_\beta\right) - 2(\mathbf{r}\cdot\mathbf{J})\delta_{\alpha\beta}\right]\frac{j_1(k_0 r)}{k_0 r} \\ +2k_0^2\int d^3\mathbf{r}\left[\begin{array}{l} 5r_\alpha r_\beta(\mathbf{r}\cdot\mathbf{J}) - \left(r_\alpha J_\beta + r_\beta J_\alpha\right)r^2 \\ -r^2(\mathbf{r}\cdot\mathbf{J})\delta_{\alpha\beta}\end{array}\right]\frac{j_3(k_0 r)}{(k_0 r)^3}\end{array}\right\}, \tag{1.47c}$$

$$Q^m_{\alpha\beta} = 15\int d^3\mathbf{r}\left\{r_\alpha(\mathbf{r}\times\mathbf{J})_\beta + r_\beta(\mathbf{r}\times\mathbf{J})_\alpha\right\}\frac{j_2(k_0 r)}{(k_0 r)^2}, \tag{1.47d}$$

where $\alpha, \beta = x, y, z$ denoting the Cartesian components of the corresponding vectors (**p**, **m**, **J**, $\boldsymbol{r}$, and their cross products) and tensors ($\mathbf{Q^e}$, $\mathbf{Q^m}$, and the Kronecker delta δ). Equation (1.47) presents the exact multipole moments valid for any particle size and shape, and they can be further related to the contributions to the total scattering cross section in a way analogous to Eq. (1.42):

$$\begin{aligned} C_{\text{sct}} = C^p_{\text{sct}} + C^m_{\text{sct}} + C^{Q^e}_{\text{sct}} + C^{Q^m}_{\text{sct}} + \ldots = \frac{k_0^4}{6\pi\varepsilon_0^2|\mathbf{E}_i|^2}\sum_\alpha|p_\alpha|^2 + \frac{k_0^4}{6\pi\varepsilon_0^2|\mathbf{E}_i|^2}\sum_\alpha\frac{|m_\alpha|^2}{c^2} \\ + \frac{k_0^6}{720\pi\varepsilon_0^2|\mathbf{E}_i|^2}\sum_{\alpha,\beta}\left|Q^e_{\alpha\beta}\right|^2 \\ + \frac{k_0^6}{720\pi\varepsilon_0^2|\mathbf{E}_i|^2}\sum_{\alpha,\beta}\frac{\left|k_0 Q^m_{\alpha\beta}\right|^2}{c^2} + \ldots. \end{aligned} \tag{1.48}$$

When the size of the particle is much smaller than the wavelength of light, approximations can be made to simplify the expressions in Eq. (1.47), resulting in

$$p_\alpha \approx -\frac{1}{i\omega}\left\{\int d^3\mathbf{r}J_\alpha + \frac{k_0^2}{10}\int d^3\mathbf{r}\left[(\mathbf{r}\cdot\mathbf{J})r_\alpha - 2r^2 J_\alpha\right]\right\}, \tag{1.49a}$$

$$m_\alpha \approx \frac{1}{2}\int d^3\mathbf{r}(\mathbf{r}\times\mathbf{J})_\alpha, \tag{1.49b}$$

$$Q^e_{\alpha\beta} \approx -\frac{1}{i\omega}\left\{\begin{array}{l} \int d^3\mathbf{r}\left[3\left(r_\beta J_\alpha + r_\alpha J_\beta\right) - 2(\mathbf{r}\cdot\mathbf{J})\delta_{\alpha\beta}\right] \\ +\frac{k_0^2}{14}\int d^3\mathbf{r}\left[\begin{array}{l} 4r_\alpha r_\beta(\mathbf{r}\cdot\mathbf{J}) - 5r^2\left(r_\alpha J_\beta + r_\beta J_\alpha\right) \\ +2r^2(\mathbf{r}\cdot\mathbf{J})\delta_{\alpha\beta}\end{array}\right]\end{array}\right\}, \tag{1.49c}$$

$$Q^m_{\alpha\beta} \approx \int d^3\mathbf{r}\left\{r_\alpha(\mathbf{r}\times\mathbf{J})_\beta + r_\beta(\mathbf{r}\times\mathbf{J})_\alpha\right\}. \tag{1.49d}$$

In some recent works, the second term in the braces on the right side of Eqs. (1.47a) and (1.49a) is associated with the so-called toroidal dipole and listed separately in a third family of toroidal multipoles [35]. In contrast to **p** and **m** possessing connections to oscillating charges and circular currents, a toroidal dipole corresponds to a poloidal current distribution. Because of the nonexistence of magnetic currents, higher-order toroidal multipoles appear only in the corresponding expressions of electric multipoles. The different treatments of toroidal multipoles, either merging them into the electric counterparts as in Mie theory or categorizing them as an independent group, manifest the difference in solving scattering problems [36]. In Mie theory, as seen in the above derivations for a sphere, the Mie coefficients a_n and b_n are determined by matching the tangential field components at the surface of the particle. The solutions do not classify the fields based on the charge-current distributions but their far-field properties such as polarization. Since an electric dipole and a toroidal dipole have identical radiation patterns, their contributions to the scattering cross section are mixed in a_n and not separable in Mie theory. The multipole expansion, on the contrary, starts with a known field (current) distribution **E** (**J**) inside the particle and rearranges it into distinct groups based on the characteristic field (current) structures. The difference between toroidal and electric multipoles in the radial component of the near field can thus be captured, providing a route to further separating them. In some emerging research topics of nanophotonics such as non-radiating excitations (i.e., anapoles), treating toroidal multipoles as a third class of elementary sources in addition to electric and magnetic ones enables a clear physical picture for interpretation and design.

1.2.2 Interactions Between a Sphere and a Dipole

Other than the Lorenz-Mie approach using the vector spherical harmonics expansion, the same formalism can be obtained by the Hertz vector (Debye potential) method [37]. Without repeating the derivation for the scattering of plane waves, in the following, we show how the interactions between a sphere and a dipolar emitter can be solved with the Hertz functions. The framework of solving the dipole-sphere coupling is sometimes referred to as generalized Mie theory[1].

The coupled dipole-sphere system has several possible combinations: the dipole by nature can be from the electric or magnetic transition of an atomic system (i.e., an electric dipole, a magnetic dipole, or a superposition of them), and it can be positioned outside or inside the sphere and be aligned tangential or vertical to the sphere's surface. The dielectric function of the sphere also has a significant impact on the interaction, but the mathematical treatment varies based only on two sets of parameters: the nature of the dipole and its orientation, as shown in Fig. 1.13. As an example, when an electric dipole **p** is tangentially coupled to a sphere of radius a at a distance d from the north pole and the origin of the coordinates is aligned to the center of the sphere, the

[1] More generally and frequently, the term "generalized Lorenz-Mie theory" is used when the incident field has a complex beam profile [54].

total fields can be expressed by two auxiliary potentials u and v, of which the magnetic and electric fields are transverse to the radial direction, respectively. In the region $r > r_s = a + d$ concerning the far field, these potentials are written as [37–39]

$$u = \frac{i|\mathbf{p}|}{r_s r}\sin\theta\cos\varphi\sum_{n=1}^{\infty}\frac{(2n+1)}{n(n+1)}\left[a_n\zeta_n'(k_0 r_s) - \varphi_n'(k_0 r_s)\right]\zeta_n(k_0 r)P_n'(\cos\theta), \tag{1.50a}$$

$$v = -\frac{|\mathbf{p}|}{r_s r}\sin\theta\sin\varphi\sum_{n=1}^{\infty}\frac{(2n+1)}{n(n+1)}\left[b_n\zeta_n(k_0 r_s) - \varphi_n(k_0 r_s)\right]\zeta_n(k_0 r)P_n'(\cos\theta). \tag{1.50b}$$

Expressions for the regions $r \leq a$ and $a < r \leq r_s$ can be found in [37]. The same as in the prior section, φ_n is the Riccati-Bessel function related to the spherical Bessel function, ζ_n is Riccati-Bessel function related to the spherical Hankel function of the first kind, and the prime denotes a derivative. The terms with Mie coefficients a_n and b_n, which are determined from boundary conditions at $r = a$ and just identical to Eq. (1.36), correspond to the scattered field, whereas the other terms in the square brackets account for the incident field from the dipole. For a vertical electric dipole, v vanishes due to unmatched symmetry, leaving the only contribution from u:

$$u = \frac{i|\mathbf{p}|}{k_0 r_s r}\sum_{n=1}^{\infty}(2n+1)\left[\varphi_n(k_0 r_s) - a_n\zeta_n(k_0 r_s)\right]\zeta_n(k_0 r)P_n(\cos\theta). \tag{1.51}$$

In both cases, the field components associated with each potential are given by

$$E_r = \frac{1}{r}\Delta^* u, \tag{1.52a}$$

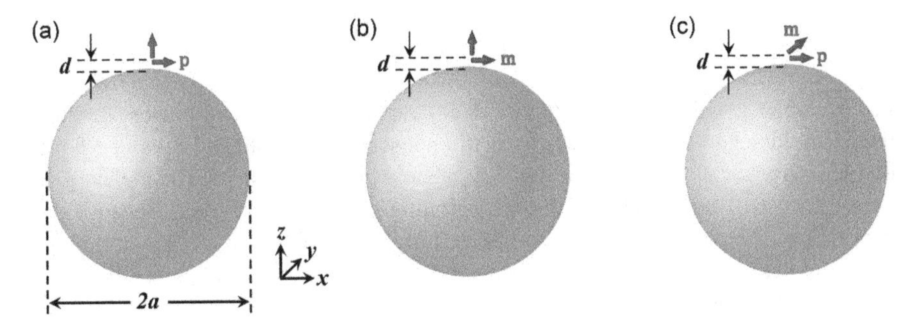

Fig. 1.13 A dipolar emitter coupled to a sphere of radius a at a distance d. The emitter can be an electric dipole $\mathbf{p}$ (**a**) or a magnetic dipole $\mathbf{m}$ (**b**), and it can be aligned tangential or vertical to the sphere's surface. (**c**) The emitter can also be chiral, containing non-orthogonal electric and magnetic dipole moments

$$E_\theta = -\frac{1}{r}\frac{\partial^2}{\partial r \partial \theta}(r \cdot u) + \frac{i\omega}{c\sin\theta}\frac{\partial v}{\partial \varphi}, \tag{1.52b}$$

$$E_\varphi = -\frac{1}{r\sin\theta}\frac{\partial^2}{\partial r \partial \varphi}(r \cdot u) - \frac{i\omega}{c}\frac{\partial v}{\partial \theta}, \tag{1.52c}$$

$$H_r = -\frac{1}{r}\Delta^* v, \tag{1.52d}$$

$$H_\theta = \frac{ick^2}{\omega\sin\theta}\frac{\partial u}{\partial \varphi} + \frac{1}{r}\frac{\partial^2}{\partial r \partial \theta}(r \cdot v), \tag{1.52e}$$

$$H_\varphi = -i\frac{ck^2}{\omega}\frac{\partial u}{\partial \theta} + \frac{1}{r\sin\theta}\frac{\partial^2}{\partial r \partial \varphi}(r \cdot v), \tag{1.52f}$$

where the operator $\Delta^* = \frac{1}{\sin\theta}\frac{\partial}{\partial\theta}\left(\sin\theta\frac{\partial}{\partial\theta}\right) + \frac{1}{\sin^2\theta}\frac{\partial^2}{\partial^2\varphi}$. The solutions for amagnetic dipole **m** can be derived in a similar manner, but a more convenient way is to take advantage of the duality theorem and replace every electric and magnetic quantities in Eq. (1.52) with their dual quantities. For more complex situations, such as a chiral emitter containing both electric and magnetic dipole moments, the solution is simply the superposition of fields contributed by **p** and **m** [40–42]. Essentially, with the analytical expression of the fields known everywhere, the interactions between the sphere and the dipolar emitter can be readily characterized.

The cross sections of a sphere coupled with a dipole can be defined following Eq. (1.42) but may not be measured by experiments as easily as in the case of the plane wave illumination. Instead, integration of the total power radiated into the far-field region is of more practical interest, as it allows one to calculate the modification of the dipole's radiative decay rate through the following relation [43]:

$$\frac{\gamma_r}{\gamma_0} = \frac{P_r}{P_0}. \tag{1.53}$$

In Eq. (1.53), γ and P are the radiative decay rate and the radiated power reaching the far-field region, respectively, with the subscript 0 denoting the values for a dipole in free space and r for the same dipole coupled to a sphere. The meaning of decay rates will be introduced in the beginning of the next chapter. Because the expression of P_0 is well formulated in electrodynamics for both electric and magnetic dipoles, the enhancement factors of the radiative decay rates for the four dipole-sphere combinations are [44–46]

$$\frac{\gamma_r^{\parallel,e}}{\gamma_0} = \frac{3}{4}\sum_{n=1}^{\infty}(2n+1)\left[\left|j_n(k_0 r) - b_n h_n^{(1)}(k_0 r)\right|^2 + \left|\frac{\varphi_n'(k_0 r) - a_n\zeta_n'(k_0 r)}{k_0 r}\right|^2\right], \tag{1.54a}$$

$$\frac{\gamma_r^{\perp,e}}{\gamma_0}=\frac{3}{2}\sum_{n=1}^{\infty}(2n+1)n(n+1)\left|\frac{j_n(k_0r)-a_nh_n^{(1)}(k_0r)}{k_0r}\right|^2, \tag{1.54b}$$

$$\frac{\gamma_r^{\parallel,m}}{\gamma_0}=\frac{3}{4}\sum_{n=1}^{\infty}(2n+1)\left[\left|j_n(k_0r)-a_nh_n^{(1)}(k_0r)\right|^2+\left|\frac{\varphi_n'(k_0r)-b_n\zeta_n'(k_0r)}{k_0r}\right|^2\right], \tag{1.54c}$$

$$\frac{\gamma_r^{\perp,m}}{\gamma_0}=\frac{3}{2}\sum_{n=1}^{\infty}(2n+1)n(n+1)\left|\frac{j_n(k_0r)-b_nh_n^{(1)}(k_0r)}{k_0r}\right|^2. \tag{1.54d}$$

Dropping the terms corresponding to the incident fields results in scattering efficiencies analogous to Eq. (1.44):

$$Q^{\parallel,e}=\frac{3}{4(k_0r_s)^2}\sum_{n=1}^{\infty}(2n+1)\left[|a_n|^2\left|\zeta_n'(k_0r_s)\right|^2+|b_n|^2\left|\zeta_n(k_0r_s)\right|^2\right], \tag{1.55a}$$

$$Q^{\perp,e}=\frac{3}{2(k_0r_s)^2}\sum_{n=1}^{\infty}n(n+1)(2n+1)|a_n|^2\left|\zeta_n(k_0r_s)\right|^2, \tag{1.55b}$$

$$Q^{\parallel,m}=\frac{3}{4(k_0r_s)^2}\sum_{n=1}^{\infty}(2n+1)\left[|a_n|^2\left|\zeta_n(k_0r_s)\right|^2+|b_n|^2\left|\zeta_n'(k_0r_s)\right|^2\right], \tag{1.55c}$$

$$Q^{\perp,m}=\frac{3}{2(k_0r_s)^2}\sum_{n=1}^{\infty}n(n+1)(2n+1)|b_n|^2\left|\zeta_n(k_0r_s)\right|^2. \tag{1.55d}$$

A noticeable difference between Eqs. (1.44) and (1.55) is that when the sphere is excited by a dipole, each Mie coefficient is multiplied by a distance-dependent weighting factor [39]. These factors provide a measure on how strong the Mie resonances are modulated by the source. Figure 1.14 presents the scattering efficiencies of a 200-nm silicon sphere under different excitation conditions in free space. The line shapes of the spectra are distinct from those in Fig. 1.12a predicted by the standard Mie theory.

Similarly, the non-radiative decay rate γ_{nr} and total decay rate γ_{tot} can be analytically solved [37]. These solutions lead to the evaluation of the Purcell factor and efficiency of the coupled system. We also leave the discussions to the next chapter with the applications of optical antennas.

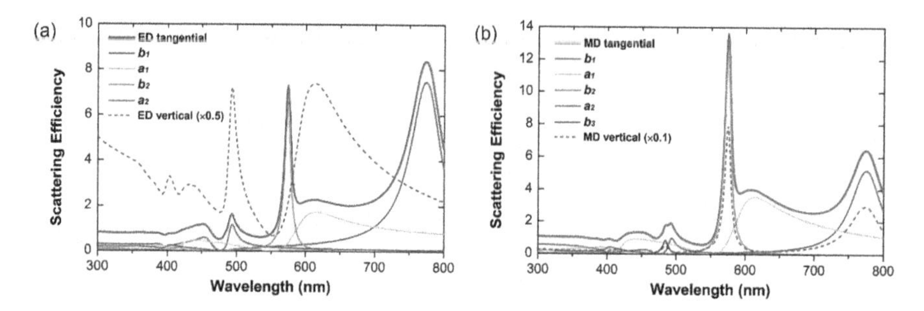

Fig. 1.14 Scattering efficiencies of a 200-nm silicon nanosphere excited by electric dipoles (**a**) and magnetic dipoles (**b**) at a distance of 1 nm. Contributions by individual Mie coefficients (solid curves) are plotted for the tangential dipoles. (Reprinted with permission from [39]. © 2021 Optica Publishing Group)

1.2.3 Scattering by a Circular Cylinder

Another important type of building blocks in nanophotonics is the nanowires. Nanowires made of high-refractive-index dielectric materials also support Mie resonances in the optical regime. In analysis, the wires are usually treated as infinitely long circular cylinders, which reduces the task to a 2D problem. The procedure of solving 2D problems is essentially akin to that introduced in Sect. 1.2.1, except the analysis is based on expansions of cylindrical harmonics. In this section, we present the formalism of normal incidence light scattering by infinite circular cylinders. For the detailed derivation and the results of oblique incidence and nanowires of a non-circular cross section or a finite length, the interested reader can refer to [11–13, 47].

Let us assume a cylinder of radius a is aligned along the z-axis in a vacuum, and the incident plane wave propagates in the $+x$ direction, as illustrated in Fig. 1.15. The cylinder is nonmagnetic and has a relative permittivity ε. Depending on the polarization of the incident wave, the solutions may take different forms. When the electric field is polarized along the axis of the cylinder (Fig. 1.15a), i.e., TE polarization, the scattering and extinction efficiencies are given by.

$$Q_{\text{sct}}^{TE} = \frac{2}{k_0 r}\left(|b_0|^2 + 2\sum_{n=1}^{\infty} |b_n|^2 \right), \tag{1.56a}$$

$$Q_{\text{ext}}^{TE} = \frac{2}{k_0 r}\text{Re}\left\{ b_0 + 2\sum_{n=1}^{\infty} b_n \right\}, \tag{1.56b}$$

with the Mie coefficients

$$b_n = \frac{J_n(kr)J_n'(k_0 r) - \sqrt{\varepsilon}J_n'(kr)J_n(k_0 r)}{J_n(kr)H_n^{(1)'}(k_0 r) - \sqrt{\varepsilon}J_n'(kr)H_n^{(1)}(k_0 r)}, \tag{1.57}$$

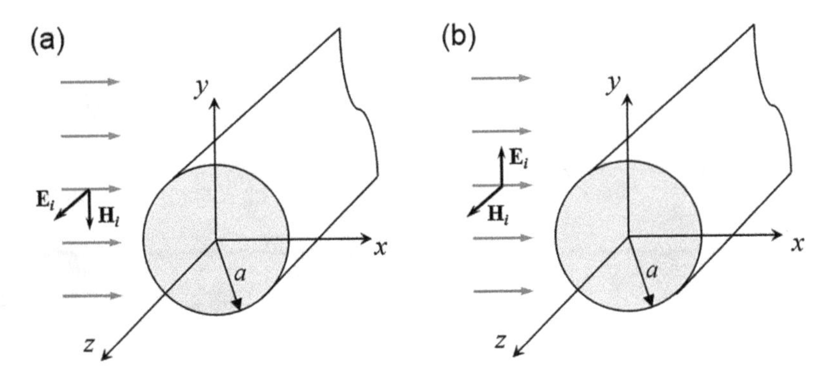

Fig. 1.15 Scattering of normal incidence plane wave by a circular cylinder of radius a. (**a**) TE polarization. (**b**) TM polarization. Conversion of Cartesian coordinates (x, y) to polar coordinates (r, θ) is defined by $r = (x^2 + y^2)^{1/2}$ and $\theta = \tan^{-1}(y/x)$

where J_n and $H_n^{(1)}$ are the Bessel and Hankel functions of the first kind, respectively. When the magnetic field is polarized along the axis of the cylinder (Fig. 1.15b), i.e., TM polarization, the efficiencies are

$$Q_{\text{sct}}^{\text{TM}} = \frac{2}{k_0 r}\left(|a_0|^2 + 2\sum_{n=1}^{\infty}|a_n|^2\right), \tag{1.58a}$$

$$Q_{\text{ext}}^{\text{TM}} = \frac{2}{k_0 r}\text{Re}\left\{a_0 + 2\sum_{n=1}^{\infty}a_n\right\}, \tag{1.58b}$$

with the Mie coefficients

$$a_n = \frac{\sqrt{\varepsilon}J_n(kr)J_n'(k_0 r) - J_n'(kr)J_n(k_0 r)}{\sqrt{\varepsilon}J_n(kr)H_n^{(1)'}(k_0 r) - J_n'(kr)H_n^{(1)}(k_0 r)}. \tag{1.59}$$

Please note that two equivalent conventions are being used in the literature to define the transverse direction, which take the cross-sectional plane and cylinder axis as the reference, respectively. Here we follow the former to keep consistency between the definitions in 2D and 3D geometries. Compared with the results in Sect. 1.2.1, despite the small difference in summation and in the leading coefficient caused by the reduced dimension, the meaning of the Mie coefficients is also different. In the 3D case, a_n and b_n coexist and correspond to electric and magnetic multipoles, respectively. For the present case, a_n is solely associated with TM-polarized plane waves, and b_n can only be excited by TE-polarized incidence (Fig. 1.16a). They can both be electric or magnetic, and the nature of each resonance needs to be determined by examining the field distributions. The alternative two-index labeling (TM_{nl} and TE_{nl}) is particularly useful in this case, since it is defined based on field patterns

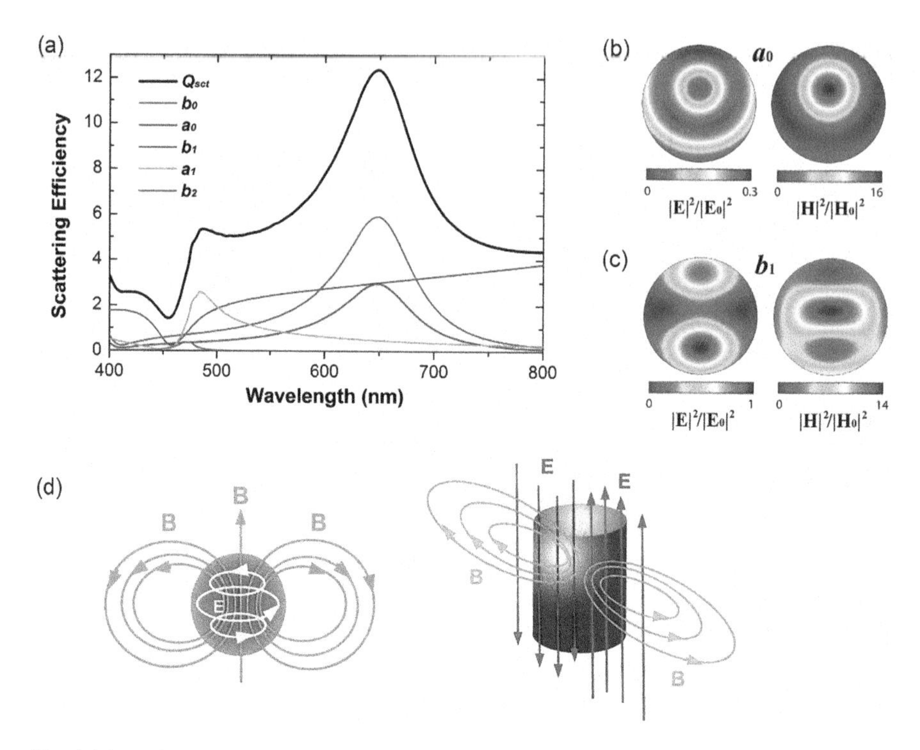

Fig. 1.16 (**a**) Scattering efficiency of a silicon nanowire of radius a = 62.5 nm. The normal incidence plane wave is unpolarized, containing TE- and TM-polarized components which are equal in magnitude. For TE (TM) polarization, only b_n (a_n) modes can be excited. (**b**, **c**) Field distributions of the nanowire in (**a**) at 650-nm wavelength, corresponding to a_0 for TM-polarized incidence (**b**) and b_1 for TE-polarized incidence (**c**), respectively. (**d**) Comparison of field structures of b_1 in a dielectric sphere and in a dielectric cylinder. (Panel (d) is reprinted with permission from [14]. Copyright (2020) Elsevier)

and can be related to the leaky modes in waveguide theory [48–50]. Figure 1.16b, c shows two examples of a_0 and b_1, which correspond to the magnetic dipole resonance under TM- and TE-polarized excitations, respectively. The field structures of b_1 in a dielectric sphere and in a dielectric cylinder are illustrated in Fig. 1.16d. Also worth mentioning is, unlike in the case of spheres, the scattering of TE-polarized plane wave off an infinite cylinder has the electric dipole (b_0) appearing at the longest wavelength, followed by magnetic dipole b_1 and higher-order modes.

The 2D version of multipole decomposition can be derived by expanding the scattered field as a series of cylindrical harmonics [51, 52]. With the configurations in Fig. 1.15, the coefficients for TE polarization are given by

$$A_m = -\frac{\eta k}{4} \int \mathbf{J}_z e^{im\varphi} J_m(kr)\, ds, \tag{1.60}$$

and those for TM polarization are

$$B_m = \frac{i}{4}\int \frac{e^{im\varphi}}{r}\left[kr\mathbf{J}_\varphi J_{m+1}(kr) + m\left(i\mathbf{J}_r - \mathbf{J}_\varphi\right)J_m(kr)\right]ds. \tag{1.61}$$

Here, m takes integers, η is the wave impedance of the host medium, k is the wavenumber, r and φ are the polar coordinates, J_m is the Bessel function of the first kind and order m, $\mathbf{J} = -i\omega\varepsilon_0(\varepsilon - \varepsilon_h)\mathbf{E}$ is the induced current density with ε_h the relative permittivity of the host medium, and the integrals are computed over the cross-sectional area of the cylinder. These coefficients are related to the scattering cross section by

$$C_{\text{sct}} = \frac{4}{k\left|\mathbf{E}_i\right|^2}\sum_{m=-\infty}^{m=\infty}\left(\left|A_m\right|^2 + \eta^2\left|B_m\right|^2\right). \tag{1.62}$$

Equation (1.62) has a form similar to that of Eqs. (1.42), (1.56a), and (1.58a), but the summation runs over all the integer orders, including negative ones. Practically, the series is truncated at a finite m. Moreover, when the scatterer is mirror symmetric and the incident plane wave propagates along the axis of symmetry, the coefficients of orders $\pm m$ are degenerate, reducing the sum to nonnegative orders only, as seen above in the case of circular cylinders.

When a nanowire is coupled to a line source parallel to its axis, Mie resonances are modulated in a similar manner as in Sect. 1.2.2 [53]. Because line sources have no physical realization in the optical region, we will not discuss this situation further.

References

1. Barnes, W.L., Dereux, A., Ebbesen, T.W.: Surface plasmon subwavelength optics. Nature. **424**(6950), 824–830 (2003)
2. Ozbay, E.: Plasmonics: merging photonics and electronics at nanoscale dimensions. Science. **311**(5758), 189–193 (2006)
3. Maier, S.A.: Plasmonics: Fundamentals and Applications. Springer, New York (2007)
4. Zayats, A.V., Smolyaninov, I.I., Maradudin, A.A.: Nano-optics of surface plasmon polaritons. Phys. Rep. **408**(3), 131–314 (2005)
5. Born, M., Wolf, E.: Principles of Optics: Electromagnetic Theory of Propagation, Interference and Diffraction of Light, 7th edn. Cambridge University Press, Cambridge (1999)
6. Johnson, P.B., Christy, R.W.: Optical constants of the noble metals. Phys. Rev. B. **6**(12), 4370–4379 (1972)
7. Raether, H.: Surface Plasmons on Smooth and Rough Surfaces and on Gratings. Springer, Berlin (1988)
8. Berini, P., De Leon, I.: Surface plasmon–polariton amplifiers and lasers. Nat. Photonics. **6**(1), 16–24 (2012)
9. Maier, S.A., Atwater, H.A.: Plasmonics: localization and guiding of electromagnetic energy in metal/dielectric structures. J. Appl. Phys. **98**(1), 011101 (2005)
10. García de Abajo, F.J.: Colloquium: light scattering by particle and hole arrays. Rev. Mod. Phys. **79**(4), 1267–1290 (2007)

11. Kerker, M.: The Scattering of Light and Other Electromagnetic Radiation. Academic Press, New York (1969)
12. Bohren, C.F., Huffman, D.R.: Absorption and Scattering of Light by Small Particles. Wiley, New York (2008)
13. van de Hulst, H.C.: Light Scattering by Small Particles. Dover Publications, New York (1981)
14. Nieto-Vesperinas, M.: 2 – Fundamentals of Mie scattering. In: Brener, I., et al. (eds.) Dielectric Metamaterials, pp. 39–72. Woodhead Publishing, Sawston (2020)
15. Khurgin, J.B., Sun, G.: Enhancement of optical properties of nanoscaled objects by metal nanoparticles. J. Opt. Soc. Am. B. **26**(12), B83–B95 (2009)
16. Ordal, M.A., et al.: Optical properties of the metals Al, Co, Cu, Au, Fe, Pb, Ni, Pd, Pt, Ag, Ti, and W in the infrared and far infrared. Appl. Opt. **22**(7), 1099–1119 (1983)
17. Jackson, J.D.: Classical Electrodynamics, 3rd edn. John Wiley & Sons, New York (1999)
18. Aizpurua, J., Hillenbrand, R.: Localized surface plasmons: basics and applications in field-enhanced spectroscopy. In: Enoch, S., Bonod, N. (eds.) Plasmonics: From Basics to Advanced Topics, pp. 151–176. Springer, Berlin/Heidelberg (2012)
19. Wiley, B.J., et al.: Synthesis and optical properties of silver Nanobars and Nanorice. Nano Lett. **7**(4), 1032–1036 (2007)
20. Schuller, J.A., et al.: Plasmonics for extreme light concentration and manipulation. Nat. Mater. **9**(3), 193–204 (2010)
21. Barnard, E.S., et al.: Spectral properties of plasmonic resonator antennas. Opt. Express. **16**(21), 16529–16537 (2008)
22. García de Abajo, F.J.: Optical excitations in electron microscopy. Rev. Mod. Phys. **82**(1), 209–275 (2010)
23. Kretschmann, E., Raether, H.: Radiative decay of non radiative surface plasmons excited by light. Zeitschrift für Naturforschung A. **23**(12), 2135–2136 (1968)
24. Otto, A.: Excitation of nonradiative surface plasma waves in silver by the method of frustrated total reflection. Zeitschrift für Physik A Hadrons and nuclei. **216**(4), 398–410 (1968)
25. Genet, C., Ebbesen, T.W.: Light in tiny holes. Nature. **445**(7123), 39–46 (2007)
26. Kuznetsov, A.I., et al.: Optically resonant dielectric nanostructures. Science. **354**(6314), aag2472 (2016)
27. Mie, G.: Beiträge zur Optik trüber Medien, speziell kolloidaler Metallösungen. Ann. Phys. **330**(3), 377–445 (1908)
28. Hergert, W., Wriedt, T.: The Mie theory: Basics and Applications. Springer, Berlin (2012)
29. Palik, E.D.: Handbook of Optical Constants of Solids, vol. 3. Academic Press, Orlando (1998)
30. Chiasera, A., et al.: Spherical whispering-gallery-mode microresonators. Laser Photonics Rev. **4**(3), 457–482 (2010)
31. Conwell, P.R., Barber, P.W., Rushforth, C.K.: Resonant spectra of dielectric spheres. J. Opt. Soc. Am. A. **1**(1), 62–67 (1984)
32. Fenollosa, R., Meseguer, F., Tymczenko, M.: Silicon colloids: from microcavities to photonic sponges. Adv. Mater. **20**(1), 95–98 (2008)
33. Alaee, R., Rockstuhl, C., Fernandez-Corbaton, I.: An electromagnetic multipole expansion beyond the long-wavelength approximation. Opt. Commun. **407**, 17–21 (2018)
34. Alaee, R., Rockstuhl, C., Fernandez-Corbaton, I.: Exact multipolar decompositions with applications in Nanophotonics. Adv Opt Mater. **7**(1), 1800783 (2019)
35. Papasimakis, N., et al.: Electromagnetic toroidal excitations in matter and free space. Nat. Mater. **15**(3), 263–271 (2016)
36. Yang, Y., Bozhevolnyi, S.I.: Nonradiating anapole states in nanophotonics: from fundamentals to applications. Nanotechnology. **30**(20), 204001 (2019)
37. Kim, Y.S., Leung, P.T., George, T.F.: Classical decay rates for molecules in the presence of a spherical surface: a complete treatment. Surf. Sci. **195**(1), 1–14 (1988)
38. Fock, V.A.: Electromagnetic Diffraction and Propagation Problems. Pergamon Press, Oxford (1965)

39. Yao, K., Zheng, Y.: Directional light emission by electric and magnetic dipoles near a nanosphere: an analytical approach based on the generalized Mie theory. Opt. Lett. **46**(2), 302–305 (2021)
40. Guzatov, D.V., Klimov, V.V., Poprukailo, N.S.: Spontaneous radiation of a chiral molecule located near a half-space of a bi-isotropic material. J. Exp. Theor. Phys. **116**(4), 531–540 (2013)
41. Zambrana-Puyalto, X., Bonod, N.: Tailoring the chirality of light emission with spherical Si-based antennas. Nanoscale. **8**(19), 10441–10452 (2016)
42. Yao, K., Zheng, Y.: Controlling the polarization of chiral dipolar emission with a spherical dielectric nanoantenna. J. Chem. Phys. **155**(22), 224110 (2021)
43. Novotny, L., Hecht, B.: Principles of Nano-Optics, 2nd edn. Cambridge University Press, Cambridge (2012)
44. Ruppin, R.: Decay of an excited molecule near a small metal sphere. J. Chem. Phys. **76**(4), 1681–1684 (1982)
45. Klimov, V., Letokhov, V.: Electric and magnetic dipole transitions of an atom in the presence of spherical dielectric interface. Laser Phys. **15**(1), 61–73 (2005)
46. Schmidt, M.K., et al.: Dielectric antennas – a suitable platform for controlling magnetic dipolar emission. Opt. Express. **20**(13), 13636–13650 (2012)
47. Balanis, C.A.: Advanced Engineering Electromagnetics. Wiley, Hoboken (2012)
48. Stratton, J.A.: Electromagnetic Theory. Wiley, Hoboken (2007)
49. Cao, L., et al.: Engineering light absorption in semiconductor nanowire devices. Nat. Mater. **8**(8), 643–647 (2009)
50. Abujetas, D.R., Paniagua-Domínguez, R., Sánchez-Gil, J.A.: Unraveling the Janus role of Mie resonances and leaky/guided modes in semiconductor nanowire absorption for enhanced light harvesting. ACS Photonics. **2**(7), 921–929 (2015)
51. Li, S.-Q., Crozier, K.B.: Origin of the anapole condition as revealed by a simple expansion beyond the toroidal multipole. Phys. Rev. B. **97**(24), 245423 (2018)
52. Johnson, S.G., et al.: Multipole-cancellation mechanism for high-Q cavities in the absence of a complete photonic band gap. Appl. Phys. Lett. **78**(22), 3388–3390 (2001)
53. Cihan, A.F., et al.: Silicon Mie resonators for highly directional light emission from monolayer MoS_2. Nat. Photonics. **12**(5), 284–290 (2018)
54. Gouesbet G., Gréhan G.: Generalized Lorenz-Mie Theories, 2nd edn. Springer International Publishing AG, Cham (2017)

Chapter 2
Nanophotonic Devices and Platforms

Abstract Understanding the optical properties of elementary structures lays the foundation of creating more complex optical phenomena, a process that largely relies on rational design. The design of nanostructures, from individual resonators to their arrangements, is one of the major activities in nanophotonics research. Depending on the degree of complexity, it can be conceived with known physical principles, derived by analytical tools, and searched through extensive simulations and optimizations. In Chap. 2, we discuss some classical nanophotonic devices, particularly from the categories of optical antennas, metamaterials, and metasurfaces. The chosen devices are exemplary of their class in illustrating the working principle, which is essential for understanding the contents in the later chapters.

2.1 Optical Antennas

Optical antennas (or, interchangeably, nanoantennas) are devices capable of converting the propagating and localized optical fields into each other [1–4]. In nanophotonics, optical antennas are an important concept for two reasons. On the one hand, analogous to the radio-frequency (RF) and microwave counterparts, optical antennas enable efficient interactions of light with nanoscale matter, such as quantum emitters and fluorescent molecules, leading to improved performance for a variety of applications including but not limited to microscopy, sensing, photovoltaics, and light sources. On the other hand, owing to their resonant characteristic, nanoantennas can strongly scatter or "reemit" light in a predefined manner and are thus used as building blocks of metasurfaces for transforming waveforms [5]. This section introduces the basic concepts of optical antennas.

The original version of this chapter was revised. The correction to this chapter is available at https://doi.org/10.1007/978-3-031-20473-9_7

K. Yao, Y. Zheng, *Nanophotonics and Machine Learning*, Springer Series in Optical Sciences 241, https://doi.org/10.1007/978-3-031-20473-9_2

2.1.1 Properties and Parameters of Optical Antennas

To give a definition to optical antennas before any discussion, we quote the statement from Ref. [1]—a compact history of antenna is also provided in the nice review that "an optical antenna is a device designed to efficiently convert free-propagating optical radiation to localized energy, and vice versa." This definition was made in the sense that the principle of optical antenna is developed based on and conceptually in analogy to that of the RF and microwave antennas, as sketched in Fig. 2.1. Based on the reciprocity theorem, the transmitting and receiving abilities of an (optical) antenna are not independent but connected.

Despite the obvious analogy, optical antennas differ from RF and microwave antennas on certain aspects. The most significant one is probably the form of the transmitters and receivers. For example, in the transmitting mode, classical antennas are fed by transmission lines that carry electric currents, whereas the sources of optical antennas are localized light emitters such as quantum dots (QDs), molecules, atoms, or ions. In both cases, the transmitters are much smaller (~1/100 in size) compared to the wavelengths of operation, but the absolute size and localized nature of the light emitters make it necessary to introduce some modifications to the well-established antenna theory for radio wave and microwave [6]. In the generic problem of optical antennas, the transmitter or receiver is usually described by an electric dipole. This classical model was proven sufficient to resemble a two-level quantum emitter that mostly resides in the ground state as well as any subwavelength neutral system of charges to the first order in multipole expansion [7]. An optical antenna, in its transmitting mode, modifies the dipole emission in several ways. To see this, we first consider the radiation properties of an oscillating electric dipole **p**.

Figure 2.2a sketches the geometry of the problem. For the convenience of analysis, the dipole **p** is aligned along the z-axis at the origin, and the fields are expressed in the spherical coordinates. Without repeating the derivation, which is available in almost all the electrodynamics and electromagnetics textbooks, the fields of such a dipole **p** in a dielectric medium of permittivity ε_d are given by

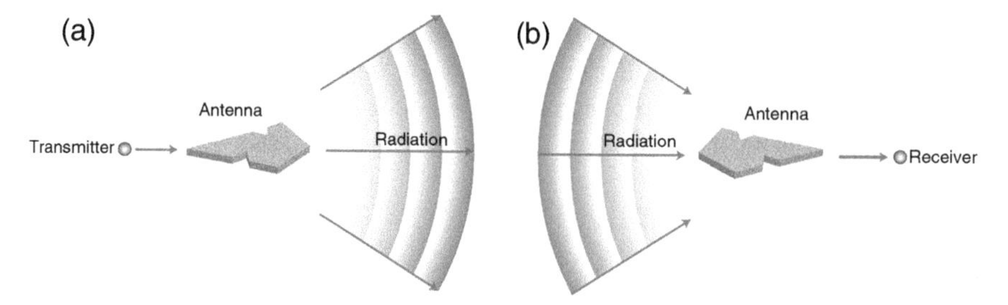

Fig. 2.1 Principle of optical antenna for transmitting (**a**) and receiving (**b**) electromagnetic energy. The transmitter or receiver is an object sized in the order of ~1/100 of the operating wavelengths, ideally a quantum emitter or absorber, such as an atom, ion, quantum dot, molecule, etc. (Reprinted from [2] with permission from Springer Nature)

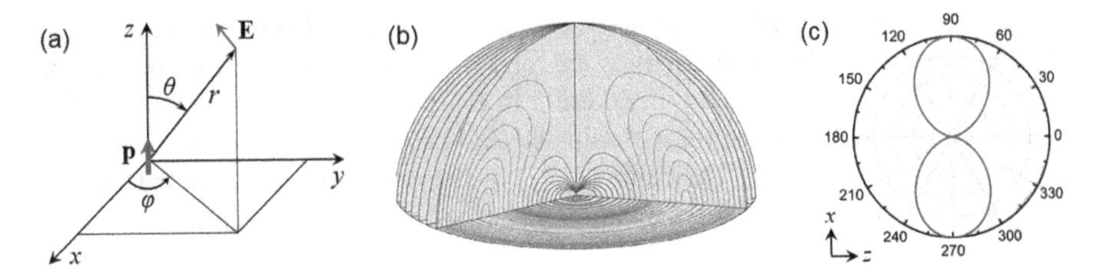

Fig. 2.2 Radiation of an oscillating electric dipole. (**a**) The dipole **p** is aligned vertical at the origin of a spherical coordinate system. (**b**) The electric field (in blue and green) and magnetic field (in red) patterns of the vertical dipole. Only the upper half-space is shown because of mirror symmetry. (c) Directional pattern of the dipole on a plane containing the z-axis. The diagram outlines the angular dependence of the radiation intensity in the radial direction. In contrast, the pattern on the x-y plane is omnidirectional. (Panel (b) is reprinted from [8] with permission. Copyright (2016) IOP)

$$E_r = \frac{|\mathbf{p}|\cos\theta}{4\pi\varepsilon_0\varepsilon_d}\frac{e^{ikr}}{r}k^2\left[\frac{2}{k^2r^2} - \frac{2i}{kr}\right], \tag{2.1a}$$

$$E_\theta = \frac{|\mathbf{p}|\sin\theta}{4\pi\varepsilon_0\varepsilon_d}\frac{e^{ikr}}{r}k^2\left[\frac{1}{k^2r^2} - \frac{i}{kr} - 1\right], \tag{2.1b}$$

$$H_\varphi = \frac{1}{Z_d}\frac{|\mathbf{p}|\sin\theta}{4\pi\varepsilon_0\varepsilon_d}\frac{e^{ikr}}{r}k^2\left[-\frac{i}{kr} - 1\right], \tag{2.1c}$$

where Z_d is the wave impedance of the medium. All the other field components are vanishing for the given system. Note that Eq. (2.1a) and (2.1b) can be directly reformulated from Eq. (1.24) by choosing $\mathbf{p} = |\mathbf{p}|\cdot\mathbf{n_z}$ with $\mathbf{n_z}$ the unit vector in the $+z$ direction. A visualization of these fields is presented in Fig. 2.2b [8]. In the far-field region where $kr \gg 1$, the electric field is dominated by its polar component E_θ, which forms with H_φ a transverse electromagnetic field propagating mainly in directions perpendicular to the dipole orientation following $\sin^2\theta$. The typical radiation pattern of a dipole **p** is thus a toroid symmetric about its axis, as shown in Fig. 2.2c.

The radiation properties of the emitter can be modified by coupling it to an optical antenna. The most obvious and straightforward change happens to the angular distribution of the radiated power. Essentially, an optical antenna "redirects" the radiation of the emitter into certain directions with concentrated power density. Such angular dependence of radiation intensity is characterized by a function $p(\theta,\varphi)$, which can be highly directional if engineered properly. By integrating $p(\theta,\varphi)$ over a surface enclosing the emitter and the antenna, we obtain the radiated power P_{rad}. This leads to the definition of a fundamental parameter of antennas, the directivity

$$D(\theta,\varphi) = \frac{4\pi}{P_{\text{rad}}}p(\theta,\varphi). \tag{2.2}$$

The first term on the right side is the reciprocal of the radiated power averaged over all directions. Therefore, directivity is the ratio of the radiation intensity in a given direction to the averaged value. In addition to P_{rad} carried by the free-space radiation, a certain amount of power is dissipated into heat and absorbed by the optical antenna, assigned to P_{loss}. The ratio of P_{rad} to the total power dissipated by the antenna P defines another useful measure, the efficiency e_{rad} of the antenna:

$$e_{\mathrm{rad}} = \frac{P_{\mathrm{rad}}}{P} = \frac{P_{\mathrm{rad}}}{P_{\mathrm{rad}} + P_{\mathrm{loss}}}. \tag{2.3}$$

It is easy to see that $e_{\mathrm{rad}} \leq 1$, and the equator holds only when the antenna elements are made of lossless materials. Combining the directivity D and efficiency e_{rad} results in the gain of the antenna:

$$G = e_{\mathrm{rad}} D = \frac{4\pi}{P} p(\theta,\varphi), \tag{2.4}$$

which takes into account both the losses of the antenna and its directional properties.

The three parameters defined above describe the characteristics of antennas in general. For optical antennas, it is worth mentioning that the term P_{loss} in Eq. (2.3) has more than one channel, namely, the absorption by the antenna. The transmitter or emitter itself also has internal losses from the non-radiative transitions, and the intrinsic quantum efficiency of the emitter is defined by

$$\eta_i = \frac{P_{\mathrm{rad}}^0}{P_{\mathrm{rad}}^0 + P_{i,\mathrm{loss}}^0}, \tag{2.5}$$

where the superscripts "0" denote the absence of the optical antenna and the "i" in the subscript indicates the internal losses of the emitter. With inclusion of the internal losses, Eq. (2.3) can be rewritten into [1].

$$e_{\mathrm{rad}} = \frac{P_{\mathrm{rad}} / P_{\mathrm{rad}}^0}{P_{\mathrm{rad}} / P_{\mathrm{rad}}^0 + P_{\mathrm{loss}} / P_{\mathrm{rad}}^0 + (1-\eta_i)/\eta_i}. \tag{2.6}$$

On many occasions, η_i is taken to be unity for the sake of simplicity of calculation, assuming the emitter is an ideal one. The total efficiency e_{rad} is therefore lower than η_i. However, in the cases where the emitters have a poor intrinsic quantum efficiency, it is possible that e_{rad} is higher. This property suggests promising applications of optical antennas in microscopy and light sources.

Another important change to an emitter coupled with an optical antenna is in the process of its spontaneous emission. A quantum emitter at an excited energy state can spontaneously transit or decay to a lower energy state, e.g., the ground state of a two-level system. This process emits a photon with the energy equal to the difference between the two states. Discovered by E. M. Purcell in 1946, the spontaneous decay rates are dependent on the dielectric environment, or specifically, the local density of electromagnetic/optical/photonic states (LDOS) [9–11]. As mentioned before, some characteristic parameters of classical antennas need to be reconsidered for optical antennas, in order to reflect their difference such as the feed source. Despite its physical significance, LDOS is a quantity in the present regard and context related to the input impedance of classical antennas defined based on circuit theory. We skip the derivation, which is available elsewhere demanding a quantum mechanical treatment, and discuss in the following from a more practical perspective. According to Fermi's golden rule, the decay rate of a quantum emitter is proportional to LDOS. Therefore, coupling an emitter with a resonator, e.g., by placing it in proximity to an optical antenna or in a cavity that has higher density of states than the free space, can increase the decay rates of the emitter, known as the Purcell effect. The enhancement of spontaneous emission is characterized by the Purcell factor F_p, defined as the ratio between the decay rate of the emitter in the coupled system, γ, and that in free space, γ_0; for $\eta_i = 1$, it can also be expressed by the ratio of the power dissipated by the coupled system, P, to the power radiated by the emitter in free space, P^0 [7]:

$$F_p = \frac{\gamma}{\gamma_0} = \frac{P}{P^0}. \tag{2.7}$$

The second equator in Eq. (2.7) relates the quantum mechanical and classical pictures of the Purcell effect. This brings great convenience in practice, because the evaluation of the spontaneous decay rates involves the dyadic Green's function, which may not be easily obtainable for complex systems, whereas the power dissipation can be computed numerically by integrating the Poynting vector over a surface that only encloses the emitter. A similar relation has been seen in Eq. (1.53), where the integration of the total radiated power P_r is defined over a surface enclosing the whole coupled system, giving the radiative part of the Purcell factor. Lastly, it should be emphasized that the interaction between a quantum emitter and a resonant nanostructure nearby modifies not only the emitter's decay rates but also the energy levels, or, in other words, the emission frequency. The latter becomes non-negligible when the interaction is strong enough to fall in the "strong coupling" regime. Strong coupling is an emerging research field in nanophotonics. Several comprehensive reviews on the theory and observation are available [12–14]. Nevertheless, our discussion only deals with the weak coupling, where the emission frequency is unaltered.

In antenna theory, the application of the reciprocity theorem helps to bridge the transmitting and receiving capabilities of an antenna. This connection holds for optical antennas as well. A useful relationship between an emitter's excitation rate γ_{exc} and spontaneous decay rate γ_{rad} can be established by [2].

$$\frac{\gamma_{\mathrm{exc}}(\theta,\varphi)}{\gamma_{\mathrm{exc}}^{0}(\theta,\varphi)}=\frac{\gamma_{\mathrm{rad}}}{\gamma_{\mathrm{rad}}^{0}}\frac{D(\theta,\varphi)}{D^{0}(\theta,\varphi)}. \tag{2.8}$$

The superscripts 0 denote variables in the absence of the antenna, and all the angle-dependent functions are defined for the same polarization state. Clearly, the enhancement of the excitation rate can be greater than that of the emission rate, if the excitation light comes from a direction of high directivity.

Optical antennas also increase the efficiency of an emitter capturing light. Equivalent to the antenna aperture of classical antennas, the absorption cross section σ of optical antennas describes the effective area that the incident light is absorbed, and it scales with the enhancement of local field intensity:

$$\sigma=\sigma_0\frac{\left|\mathbf{n}_p\cdot\mathbf{E}\right|^2}{\left|\mathbf{n}_p\cdot\mathbf{E}_0\right|^2}. \tag{2.9}$$

Here, σ_0 is the absorption cross section of the dipolar emitter $\mathbf{p}$ without being coupled to an antenna, $\mathbf{n}_p$ is the unit vector along the dipole orientation, and $\mathbf{E}$ and $\mathbf{E}_0$ are the electric fields at the emitter in the presence and absence of the antenna, respectively. Given that the local field enhancement by a resonant nanostructure can be as strong as 2–3 orders of magnitude, an emitter that is coupled to an optical antenna effectively interacts with the incident light in an area about 10^4–10^6 times of its real spatial extension.

Before proceeding further to specific examples, we summarize in Fig. 2.3 some typical designs of optical antenna. As can be seen, the number, shape, and arrangement of the antenna elements can be very different. These variations are aimed at enhancing different aspects of the antenna's properties and performance, such as directivity and LDOS, providing a rich toolbox to manipulate light-matter interactions on the nanometer scale.

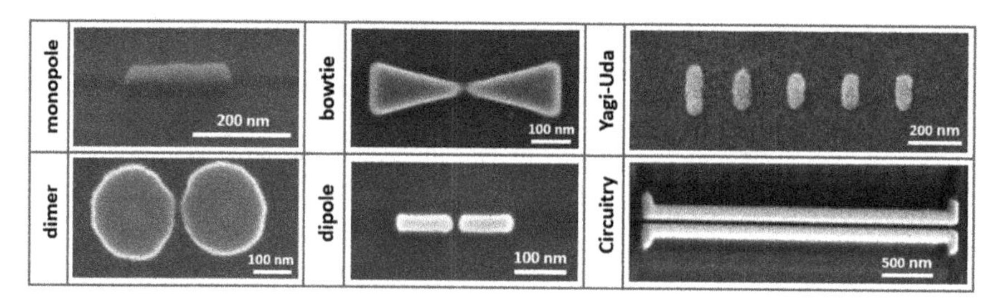

Fig. 2.3 Scanning electron microscope (SEM) images of optical antennas (top to bottom, left to right): monopole antenna, dimer antenna, bowtie antenna, dipole antenna, Yagi-Uda antenna, and antenna circuit. Other variants include cross antennas, spiral antennas, split-ring resonators, slot antennas, etc. The devices can be made of plasmonic or dielectric materials. (Reprinted from [2] with permission from Springer Nature, from [3] with permission from IOP, and from [15] with permission from Springer Nature, respectively)

2.1.2 Single Particles as Optical Antennas

Individual nanoparticles can function as optical antennas. The simplest form of these monopole antennas is a sphere. Because the directional behavior of optical antennas is a major characteristic concerned in antenna design and engineering, we mainly consider the radiation patterns herein and leave the brief discussion on spontaneous emission to Sect. 2.1.4. Also, we would extend the applicable range of the term "nanoantennas" slightly to include resonant nanoparticles that have exotic scattering properties but are not necessarily coupled with localized emitters. With no disruption to introducing the basic principle, this extension allows for a smooth transition as we move on to metasurfaces in the next section.

As discussed above, the radiation of an isolated dipole is omnidirectional on the plane perpendicular to the dipole orientation. The symmetry of the radiation pattern can be broken by placing a plasmonic nanosphere near the emitter in the configuration sketched in Fig. 2.4a. The particle either reflects or attracts the dipole emission, depending on several variables, such as the dipole-sphere distance [16, 17], the wavelength of operation, and the size of the sphere [18], as summarized in Fig. 2.4b, c. Although the directional behaviors of this system can be fully assessed with rigorous analysis by methods like generalized Mie theory, it is instructive to examine them using a simple yet physically intuitive model, which will help us to understand more sophisticated cases where analytical solutions do not exist. Recalling LSPs in Sect. 1.1.2, under the quasi-static approximation, an electric dipole moment in the particle will be induced by the external field. For the present system, if we simplify the plasmonic nanosphere into an induced dipole $\mathbf{p}_{\text{in}}$ at its center, the problem reduces to solving the interference of two dipoles: an exciting dipole $\mathbf{p}$ and an induced dipole $\mathbf{p}_{\text{in}}$ at a distance d. The latter can be calculated by using Eq. (1.26), with the electric field $\mathbf{E}_0$ replaced by the exciting dipole field at the center of the sphere. Therefore, the relative phase between the two dipoles is determined by their distance and the polarizability of the sphere. Strictly speaking, this two-dipole model may be oversimplified. Compared with the exact solution, the treatment here ignores the fact that the nanosphere is in the near field of the exciting dipole, which varies rapidly across the dimension of the particle, inducing higher-order resonances. However, because of the dominance of the dipolar mode in radiation, the seemingly naive model predicts the far-field characteristics quite well. As an example, Fig. 2.4b compares the reflected power efficiencies of an 80-nm silver sphere calculated with dipole approximation and by multipole expansion. The reflected power efficiency is defined as the ratio of the power radiated into the half-space where $x < 0$ (see Fig. 2.4a) to the total power radiated in all directions. Only slight deviations are seen for small dipole-sphere distances, confirming the validity of the two-dipole-interference model in explaining the far-field radiation properties. On the other hand, higher-order modes make significant contributions to the decay rates and cannot be simply neglected when evaluating near-field quantities.

The picture of interference can be extended to explain the directional behavior of antennas involving multipoles or dipole arrays. Let us first consider the cases of

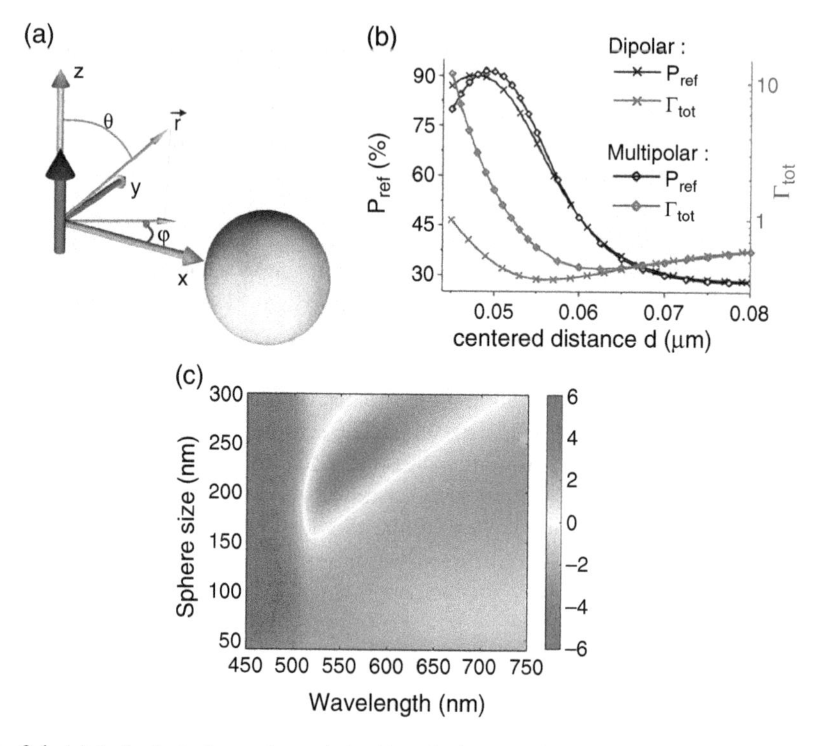

Fig. 2.4 (**a**) A dipole (red arrow) coupled with a single nanosphere antenna. (**b**) Dependence of the reflected power efficiency on dipole-sphere distance at the emission wavelength of 600 nm. The sphere is 80 nm in diameter and made of silver, and the refractive index of the surrounding medium is 1.5. (**c**) Forward-to-backward emission ratio as a function of wavelength and sphere size, when an electric dipole is tangentially coupled with a gold sphere at 1 nm in free space. (Panels (a) and (b) are reprinted from [16] with permission. © 2011 Optica Publishing Group)

multipoles, which could arise from a variety of single nanoparticles, such as high-refractive-index dielectric beads and specially shaped metallic resonators [19]. To illustrate how the interference between multipoles can result in directional scattering, an analysis based on the phase symmetry of multipoles is employed [20]. Figure 2.5a shows the principle of suppressing backward scattering with the lowest four orders of multipoles in a dielectric sphere, namely, the electric dipole (ED), magnetic dipole (MD), electric quadrupole (EQ), and magnetic quadrupole (MQ) (see Fig. 1.11), all excited by a plane wave propagating from the left with in-plane electric fields. With this choice of polarization, ED and EQ are in-plane, whereas MD and MQ are out-of-plane. The corresponding in-plane and out-of-plane far-field scattering patterns are summarized and color-coded in the middle row of Fig. 2.5a. The phase symmetries of the far field of individual multipoles comply with the following rule: the electric fields have opposite parities for (i) multipoles of the same nature and adjacent orders and (ii) multipoles of different natures but the same order. The former case applies to, e.g., the ED and EQ pair. The electric field (red arrows) of an ED has even symmetry, while the electric field of an EQ has odd

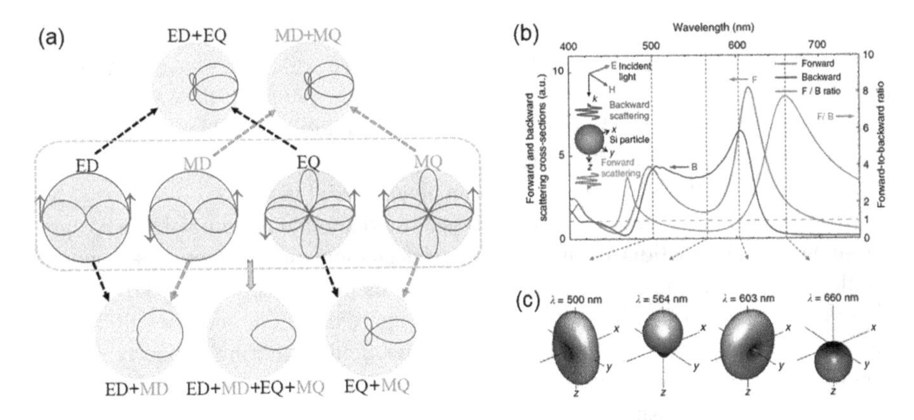

Fig. 2.5 (**a**) Phase-symmetry analysis for directional scattering based on interference of multipoles. With proper alignments and balanced amplitudes, electric dipole, magnetic dipole, electric quadrupole, and magnetic quadrupole are paired up or mixed altogether to produce directional scattering. Curves in purple and blue represent the in-plane and out-of-plane scattering patterns, respectively. Red arrows indicate the relative phase of the electric fields in the forward and backward directions. (**b, c**) Directional scattering of light by a silicon nanosphere with a radius of 75 nm in free space. (**b**) Forward and backward scattering cross sections and their ratio. (**c**) Far-field scattering patterns at four selected wavelengths. (Panel (a) is reprinted from [20] with permission. © 2018 Optica Publishing Group. Panels (b) and (c) are reprinted from [21] with permission from Springer Nature)

symmetry. When their amplitudes are balanced and phases are properly aligned, ED and EQ emit light in such a way that their fields interfere constructively in the forward direction and destructively in the backward direction, leading to a directional scattering pattern with suppressed backscattering (top left panel). A similar analysis can be deduced for MD and MQ.

The latter case has a representative example of the ED and MD pair. Owing to their opposite phase parities, total cancellation of backward (bottom left panel) or forward scattering is achievable with properly overlapped ED and MD. These effects were discovered by Kerker et al. [22], and the conditions for zero-backward and zero-forward scattering are named after him as the first and second Kerker conditions [23], respectively. Antennas that operate in this principle are known as Huygens sources [24]. The high directivity of these elementary sources holds the promise for controlling the reflectance and transmittance of metasurfaces after integration. Observation of Kerker-type directional scattering by high-refractive-index subwavelength spheres has been reported by several groups [21, 25, 26]. Figure 2.5b, c shows one example for silicon nanoparticles, where the maximum forward-to-backward scattering ratio is obtained at ~660-nm wavelength.

The analysis remains valid for more sophisticated situations, where, for example, higher-order modes also come into play, or, the multipoles are excited by a local emitter. Involvement of higher-order multipoles could help to improve directivity, as shown in the bottom row of Fig. 2.5a. In fact, the constructive or destructive role of each multipole in determining backscattering can be recognized by inspecting

Eq. (1.46), where the opposite signs of a_n and b_n and the alternating sign of the n-th order terms just coincide with the aforementioned rule on phase symmetry. Excitation by local emitters offers another variable, the nature of the emitter, to the interference process. It has been demonstrated that at close wavelengths, the emission of an electric dipolar emitter and of a magnetic dipolar emitter can be directed into opposite directions by the same silicon nanosphere [18].

We then consider nanorods, another class of fundamental building blocks of optical antennas. To function as an antenna or an element of an antenna, a nanorod should carry an electric current oscillating along its main axis (see Fig. 1.7a for the longitudinal resonance). Under the quasi-static approximation, one can still use the induced dipole moment $\mathbf{p}_{\text{in}}$ to describe the far-field scattering property of the nanorod; the only difference from the case of a sphere is that the polarizability is now a function of the aspect ratio.

As the length of the nanorod increases, the quasi-static approximation gradually fails, and some other methods should be considered for analysis. By treating highly elongated nanorods as nanowires, the Fabry-Pérot model has proved both effective and intuitive [28, 29], while several more accurate and sophisticated analytical models are also available [27, 30]. Figure 2.6a, b summarizes the comparison between simulated and analytically solved near fields of two gold nanowires at the first- and third-order resonances, respectively. Experimentally, these patterns can be imaged by using NSOM. The formation of the node patterns is similar to those in Fig. 1.8 for metal strips, except the involved modes are not short-range SPPs but longitudinal resonances of the nanowires, which are essentially discretized 1D SPPs because of the finite wire lengths. On an infinitely long metal wire, the SPPs are bound at the surface, propagating along the axis with $k_{\text{sp}} > k_0$, or, equivalently, $\lambda_{\text{sp}} < \lambda_0$. The truncation of the wire leads to radiation, and a nontrivial phase shift resulted from the reflection of SPPs at both ends of the wire further modifies the radiation patterns of high-order resonances, producing noticeable differences as compared to the classical RF antenna theory [3], as shown in Fig. 2.6c.

2.1.3 Multiparticle Nanoantennas

Multiple nanoparticles that are properly arranged provide more opportunities for engineering the near- and far-field properties. Near-field distributions around a nanoantenna largely determine its functionality. As introduced in the preceding subsection, the enhancement of local field intensity is associated with the antenna aperture (or absorption cross section) as well as impedance (or LDOS). Therefore, tremendous efforts have been devoted to designing and optimizing structures which can offer stronger field enhancement, and introducing local shape singularities, such as a gap between resonant elements, turns out to be an effective strategy for this purpose. Figure 2.7 compares the near-field distributions of a nanorod antenna and of a gap antenna [31]. Under the same excitation conditions, it is obvious that the

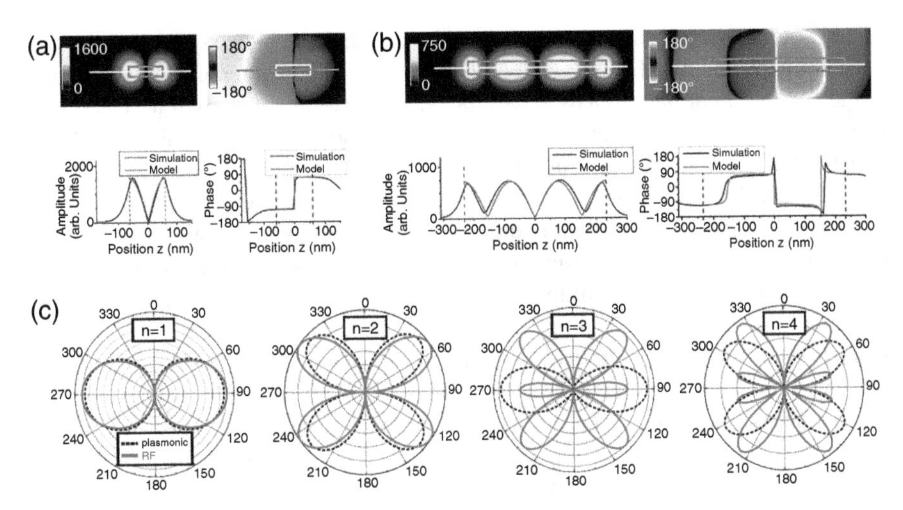

Fig. 2.6 Plasmonic nanowire antennas. Comparison between simulated and analytically solved near fields of the first-order (**a**) and third-order (**b**) resonance of two gold nanowires. Upper row: Top view of the near-field distributions on a plane slightly above the nanowire. Left/right panels are the amplitude/phase. Rectangles outline the nanowire, and the horizontal lines indicate the cutting lines. Lower row: Probed field amplitudes and phases along the cutting lines. Curves in black and red are based on simulations and an analytical model, respectively. (**c**) Comparison between the emission patterns of plasmonic (black dashed curves) and RF (red curves) wire antennas for the first four resonances. (Panels (a) and (b) are reprinted from [27] with permission. Copyright (2010) American Chemical Society. Panel (c) is reprinted from [3] with permission from IOP)

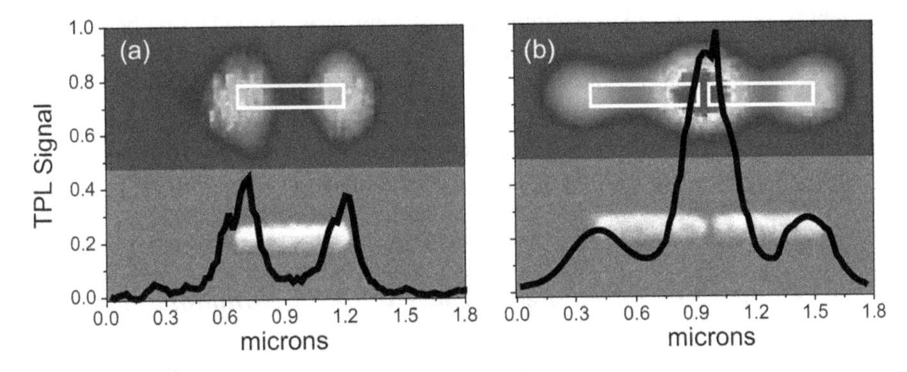

Fig. 2.7 Two-photon-induced luminescence (TPL) maps (top) and SEM micrographs (bottom) of a gold nanorod (**a**) and of a gap antenna composed of two nanorods aligned end to end (**b**), both excited at 730-nm wavelength. The TPL signal is proportional to $|\mathbf{E}|^4$. The length of the rods is 500 nm. (Reprinted from [31] with permission. Copyright (2008) American Physical Society)

field intensity is drastically enhanced at the gap, giving rise to a preferable site for attachment of nanoloads.

One difference worth mentioning between optical gap antennas and their RF counterparts is the property of the gap [1]. While in both cases the gap functions as a feed point, for RF and microwave dipole antennas, the two arms are connected to a feed source via impedance-matched feed lines, and the current through the gap does not carry localized energy. In contrast, an optical gap antenna is fed by a load in the gap region of high LDOS to facilitate the excitation and emission. In addition to nanorods, nanospheres and nanotriangles are also common building blocks in constructing gaps in a dimer configuration (see Fig. 2.3). The influence of the geometry of individual elements on antenna properties, mainly the field intensity, resonance wavelength, and radiation efficiency, has been widely studied. Detailed discussions can be found in several comprehensive reviews, such as [1–4, 32].

The coupling between multiple resonant elements is more influential in determining the far-field properties of the nanoantennas. The best example of this application is probably the optical Yagi-Uda antenna [33]. Well-known for its unidirectional radiation characteristic, a generic RF Yagi-Uda antenna, named after the two inventors, consists of a series of parallel metal rods in a linear array [6]. One of the rods in the middle is essentially a dipole antenna, serving as a driven element. The other rods are all parasitic, and the typical arrangement is to place one "reflector," i.e., a rod slightly longer than the driven element, on one side of it opposite to the radiation direction while several "directors," i.e., rods slightly shorter than the driven element, on the other side coincident with the radiation direction. The driven element induces a current in each of the parasitic rods, making them secondary sources of radiation. Therefore, the behavior of a Yagi-Uda antenna can be represented by a linear array of parallel dipoles. At optical frequencies, this design can be carried out by plasmonic nanorods and dielectric nanoparticles [34], as illustrated in Fig. 2.8a, b, with a dipole acting as the driven element. In practice, the exciting dipole has different forms of realization, such as placing a QD at one end of the feed element (see Fig. 2.8c) [35], exciting the feed element with an electron beam [36] or a light beam [37–39], or applying a DC voltage over a nanoscale feed gap [40].

The analysis of optical Yagi-Uda antennas consisting of dielectric elements is a bit more complicated due to the possible coexistence of electric and magnetic resonances. Nevertheless, the low-loss property and more tuning factors in design still make them promising solutions to achieving highly directional radiation (Fig. 2.8d) [41].

2.1.4 *Controlling Spontaneous Emission with Nanoantennas*

In Sect. 1.2.2, we have presented the formulae to compute the modification of a dipole's radiative decay rate by a coupled sphere. Because the analytical solution contains the exact field information at every point in space, integration of the Poynting vector over a surface enclosing only the emitter gives the total power dissipation P by the antenna and, in turn, the closed-form expressions of Purcell factors via Eq. (2.7) [42]:

$$\frac{\gamma_{\text{tot}}^{\parallel,e}}{\gamma_0} = 1 - \frac{3}{4}\,\text{Re}\sum_{n=1}^{\infty}(2n+1)\left\{a_n\left[\frac{\zeta_n'(k_0 r)}{k_0 r}\right]^2 + b_n\left[h_n^{(1)}(k_0 r)\right]^2\right\}, \tag{2.10a}$$

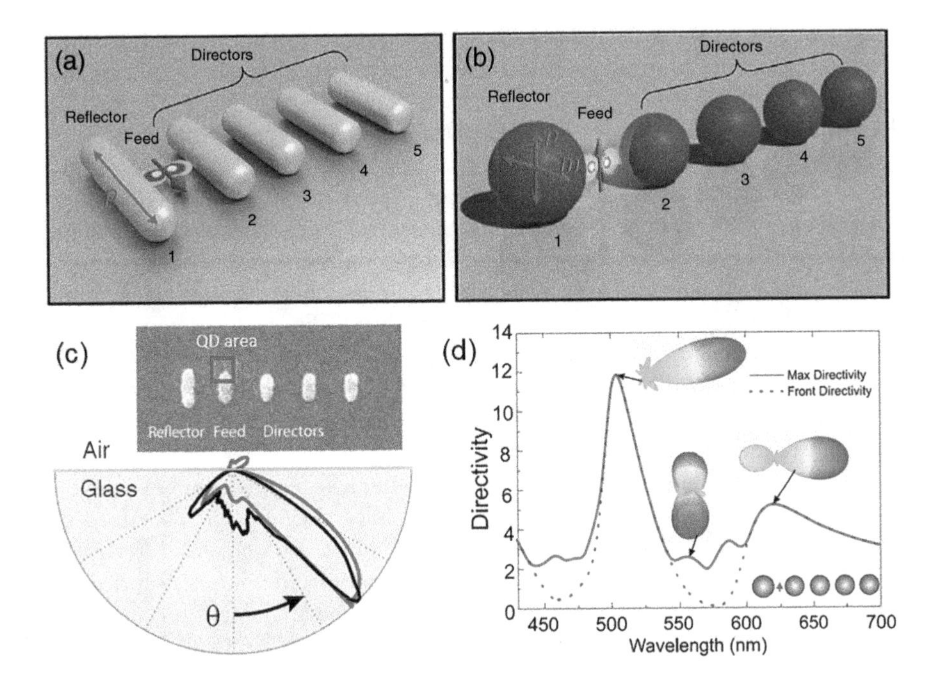

Fig. 2.8. (**a, b**) Optical Yagi-Uda antennas composed of plasmonic nanorods (**a**) and high-refractive-index dielectric nanospheres (**b**). (**c**) Radiation patterns of a five-element gold Yagi-Uda antenna, excited by a QD at one end of the feed element. (**d**) Directivity of the silicon Yagi-Uda antenna in (**b**). (Panels (a) and (b) are reprinted from [34] with permission. Copyright (2020) Elsevier. Panel (c) is reprinted from [35] with permission from AAAS. Panel (d) is reprinted from [41] with permission. © 2012 Optica Publishing Group)

$$\frac{\gamma_{\text{tot}}^{\perp,e}}{\gamma_0} = 1 - \frac{3}{2}\text{Re}\sum_{n=1}^{\infty}(2n+1)n(n+1)a_n\left[\frac{h_n^{(1)}(k_0 r)}{k_0 r}\right]^2, \tag{2.10b}$$

$$\frac{\gamma_{\text{tot}}^{\parallel,m}}{\gamma_0} = 1 - \frac{3}{4}\text{Re}\sum_{n=1}^{\infty}(2n+1)\left\{a_n\left[h_n^{(1)}(k_0 r)\right]^2 + b_n\left[\frac{\zeta_n'(k_0 r)}{k_0 r}\right]^2\right\}, \tag{2.10c}$$

$$\frac{\gamma_{\text{tot}}^{\perp,m}}{\gamma_0} = 1 - \frac{3}{2}\text{Re}\sum_{n=1}^{\infty}(2n+1)n(n+1)b_n\left[\frac{h_n^{(1)}(k_0 r)}{k_0 r}\right]^2. \tag{2.10d}$$

Here, Re(·) represents the real part of the summation in the parenthesis. The results for non-radiative decay rates can be obtained by subtracting the radiative decay rates from the total decay rates, and the efficiencies of the coupled system as an optical antenna are simply the ratios of the radiative decay rates to the total decay rates. Figure 2.9a shows the decay rates and quantum efficiencies of an electric dipole near a silver nanosphere [43]. It can be seen that, as the dipole-sphere distance decreases, the radiative decay increases with a relatively smooth slope and is dominated by the dipole mode of the sphere (upper panel), consistent with the previous model used for Fig. 2.4b. In contrast, the non-radiative decay rate increases in a much more drastic manner at small distances, where the contributions by higher-order resonances become significant (middle panel). This explains the failure of the dipole model in approximating the Purcell factor. The distinct behaviors of the radiative and non-radiative decay rates result in a non-monotonic line shape of the quantum efficiency (lower panel). On the one hand, due to the dominant non-radiative decay channel, the quantum efficiency decreases rapidly when the emitter is very close to the particle, known as quenching. On the other hand, as formulated in Eq. (2.6), for emitters that have a poor intrinsic quantum efficiency, the extrinsic efficiency of the coupled system can be substantially higher.

The spontaneous emission of emitters coupled to a sphere has been extensively studied in the literature using the analytical tool. Interesting enough, the effects on electric and magnetic dipoles are quite different. Figure 2.9b compares the extrinsic quantum efficiencies of a magnetic dipole and of an electric dipole near a silver nanosphere, suggesting that the quenching of the magnetic dipole is much weaker

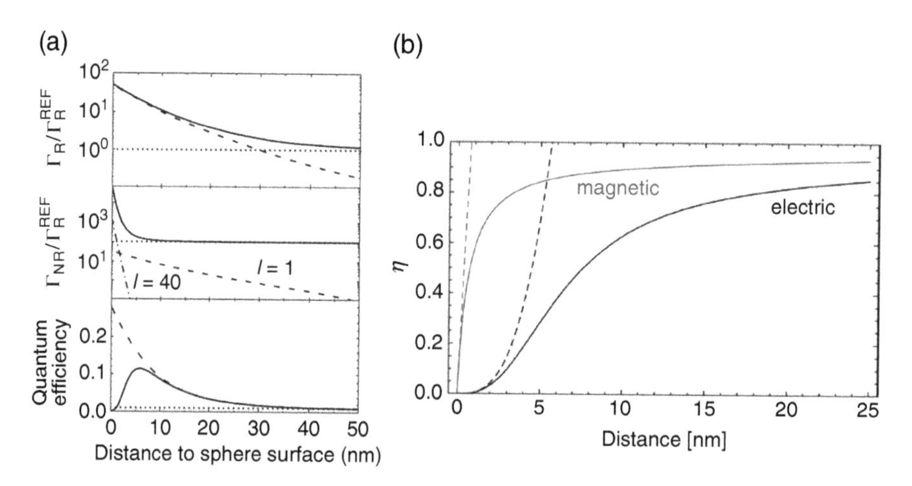

Fig. 2.9 Decay rates and efficiencies of dipole emitters coupled to a plasmonic nanoparticle. (**a**) Radiative decay rates (upper panel), non-radiative decay rates (middle panel), and extrinsic quantum efficiencies (lower panel) of an electric dipole near a silver sphere with a radius of 30 nm. The dipole has an intrinsic quantum efficiency $\eta_i = 1\%$ and an emission wavelength of 433 nm. The refractive index of the environment is 1.3. Solid lines are based on the exact solutions, dashed lines are contributions by the dipole mode unless labeled otherwise, and dotted lines are evaluated in the absence of the particle. (**b**) Comparison between the extrinsic quantum efficiencies of a magnetic dipole and of an electric dipole near a silver sphere with a radius 50 nm. Both emitters are ideal with a 100% intrinsic quantum efficiency. The refractive index of the surrounding medium is 1.5, and the wavelength is 576 nm. All the values are averaged over all orientations of the emitters. (Panel (a) is reprinted from [43] with permission. Copyright (2011) American Physical Society. Panel (b) is reprinted from [44] with permission. Copyright (2016) American Chemical Society)

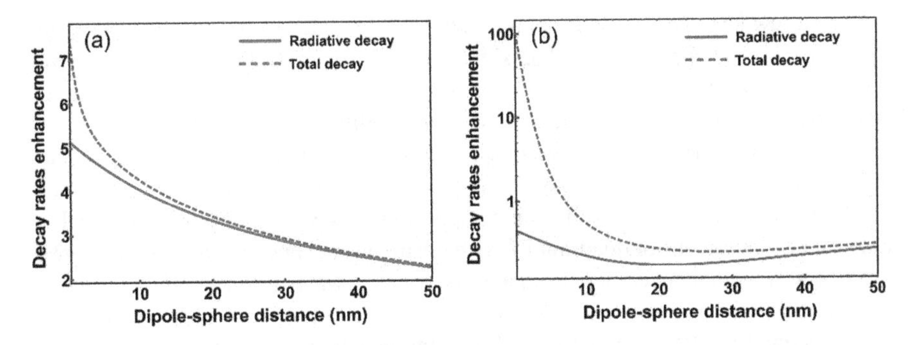

Fig. 2.10 Comparison between the Purcell factors and radiative decay rate enhancements of a magnetic dipole (**a**) and an electric dipole (**b**) coupled to a silicon sphere with a radius of 85 nm in free space. The x-axis for dipole-sphere distance is truncated at 1 nm on the left. (Reprinted with permission from [18]. © 2021 Optica Publishing Group)

than that of the electric dipole [44]. Another comparison is made for dielectric nanoparticles [18], where similar results are attained, as shown in Fig. 2.10. Extended discussions on different types of emitters and dielectric environments are available in several studies; see, e.g., [45–50].

Lastly, we remind the reader of another useful formula in evaluating the modification of spontaneous decay rates. According to Poynting's theorem, in a linear, lossless, and nondispersive medium, the time-averaged power P radiated by an emitter with a harmonic time dependence must be equal to its rate of energy dissipation. For an electric dipole $\mathbf{p}$ at $\mathbf{r}_0$, this yields [7]

$$P = \frac{\omega}{2} \mathrm{Im}\left\{\mathbf{p}^* \cdot \mathbf{E}\left(\mathbf{r}_0\right)\right\}. \tag{2.11}$$

One can further decompose the field $\mathbf{E}$ into the dipole's own field $\mathbf{E}_0$ and the field $\mathbf{E}_s$ scattered by the nanostructures, i.e., $\mathbf{E} = \mathbf{E}_0 + \mathbf{E}_s$. These two terms are connected to P_0 and the change of radiated power. When normalized by P_0 as in Eq. (2.7), the first term reduces to unity (see Eq. (2.10)). The same results apply for magnetic dipoles with the interchange of dual quantities.

2.2 Metamaterials and Metasurfaces

Another major playground of nanophotonic research is the field of metamaterials and their 2D counterparts, metasurfaces. The term "metamaterial" was coined around 2000 to name artificially constructed materials which possess novel electromagnetic responses that do not occur in nature [51–53]. Albeit this description subtly underrates the possible properties of naturally occurring materials and does not include a few key characteristics emphasized in some later definitions, it highlights the major aim of the research. The significance of enabling new electromagnetic properties can be seen in Fig. 2.11a, which classifies materials in the parameter space based on their permittivity ε and permeability μ. Noticeably, conventional transparent dielectrics only fill the first quadrant featuring positive ε and μ. Metals at frequencies lower than their plasma frequency (see Eq. (1.7)) have negative ε and positive μ, a combination that does not support propagating waves, as illustrated in quadrant II. Inclusions of quadrant IV with positive ε and negative μ are rare; known examples are magnetized ferrite materials at microwave frequencies, and waves therein must still be evanescent. Another region where propagating waves are supported is quadrant III. With both ε and μ taking negative values, a material exhibits a negative refractive index, opposite in sign to the indices of common dielectrics in quadrant I. Many exotic phenomena associated with negative refractive index were predicted, as early as in 1968 by Veselago [54], but no material in nature fulfills the requirements of simultaneously negative ε and μ. Moreover, most media dealt with in optics are nonmagnetic (i.e., μ is unity), limiting the available choice of materials to a very narrow strip denoted by the green dashed line in Fig. 2.11a.

The research of metamaterials is largely related to the pursuit of access to the traditionally unreachable regions in the electromagnetic parameter space. The key concepts allowing the solution of metamaterials to be distinct from and go "beyond" (which is what the prefix "meta" means) conventional materials are as follows [51]: metamaterials are composed of artificially structured and arranged elements with the scales of inhomogeneity much smaller than the wavelength of interest, and the properties of metamaterials are essentially determined by the elementary units rather than the constituent materials. The response of a metamaterial is therefore expressed by effective, homogenized parameters.

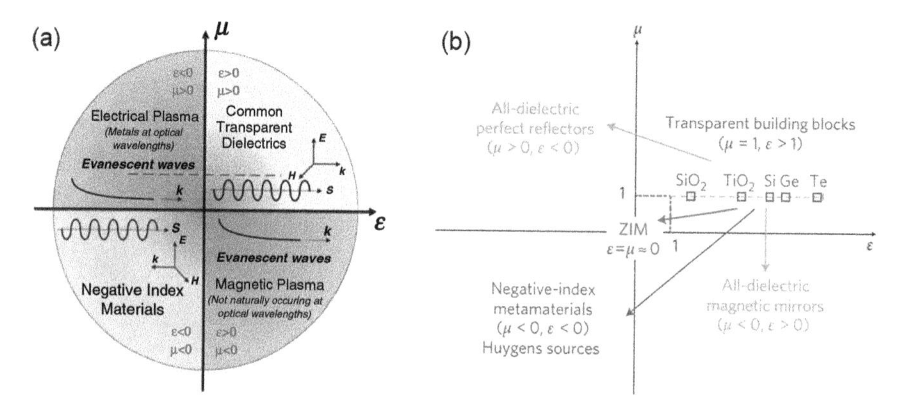

Fig. 2.11 Electromagnetic parameter space for the permittivity ε and permeability μ of (**a**) materials in general and (**b**) the subset of all-dielectric metamaterials. The horizontal and vertical axes denote the real parts of ε and μ, respectively. (Panel (a) is reprinted from [51] with permission from Springer Nature. Panel (b) is reprinted from [55] with permission from Springer Nature)

Whereas optical metamaterials manifest their unusual properties based on accumulated effects from light interacting with the artificial elements during propagation, metasurfaces achieve novel control over the wavefront of light by introducing abrupt changes on an interface, usually in the form of a single layer of optical antennas with subwavelength lateral spacing and spatially varying geometric parameters [5, 56]. In this regard, metasurfaces can be considered as a class of metamaterials with a reduced dimensionality. The invention of metasurfaces opens the door to "flat optics," where revolutionary research is ongoing.

In this section, we discuss the basics of metamaterials and metasurfaces, from the design of elementary building blocks to how they can be assembled to produce novel optical phenomena. This completes the preparation of essential knowledge of nanophotonics for the later discussion on applications.

2.2.1 Optically Resonant Building Blocks

As stated above, the lack of magnetic response in materials, especially at optical frequencies, is a major obstacle to accessing larger areas in the parameter space for novel phenomena. The reason behind this problem is that the magnetic component of an optical field interacts with atoms in a much weaker strength than the electric component does. With metamaterials, nonetheless, such limitation can be lifted, because the functional building blocks of a metamaterial are the artificially structured subwavelength elements, not any more the atoms of the constituent materials. During the early days of metamaterials, considerable efforts were devoted to the realization of artificial magnetism, or more specifically, negative permeability [57]. A variety of structures based on metals were proposed to induce strong magnetic responses to the incident magnetic field under resonant conditions. Some of them turned out to be more efficient after experimental verifications in the microwave regime, which were refined and adapted in the succeeding development for optical magnetism [58]. Because it is in general much easier to obtain the electric elements of metamaterials, we focus our discussion on the design of magnetic elements. And another note to keep in mind is that the Mie resonances introduced in Chap. 1 can be electric- or magnetic-type for high-refractive-index dielectric nanoparticles. In other words, magnetic responses are "built-in" and more dependent on the size, rather than shape, of the particles. Therefore, despite that a decent coverage of the parameter space can be achieved by using all-dielectric metamaterials (Fig. 2.11b) [55], only the design of plasmonic elements will be discussed in the following.

The elementary unit of magnetic responses is a magnetic dipole. Knowing that the absence of optical magnetism in nature results from the very weak coupling of atoms and the magnetic field of light, all about the trick needed for designing an optically active magnetic element is to induce a resonance that mimics a magnetic dipole, i.e., a small current loop. The first yet most successful design of such "meta-atoms" in metamaterial research is the split-ring resonator (SRR) [59, 60]. Figure 2.12a shows the typical geometry of a single-slit SRR, which is a square ring with an opening cut through one edge. Another popular shape of SRRs is a circular split ring. The working principle of an SRR can be intuitively understood from its geometry, as a metallic ring is known to be able to accommodate a loop of current. A better model to gain insight into the principle is the equivalent LC circuit [51], as illustrated in the inset of Fig. 2.12a. The inductance L is related to the current loop and thus can be estimated based on an SRR's size and shape. However, a complete ring only couples to the external magnetic field weakly, meaning that the inductive response alone is nonresonant. Now, it is easier to see why the ring needs to be "split." By intentionally introducing a gap, the loop of conduction current is dammed, and charges accumulate at the two opposing facets across the gap, forming a capacitor with an equivalent capacitance C connected in series to the inductor. Thus, the resonant behavior of SRRs can be well explained by the LC circuit model.

The original design of SRRs as elements of metamaterials, proposed by Pendry and coworkers, comprises a pair of split rings, one of which is slightly smaller in size and placed concentrically inside the larger ring with their gaps facing opposite directions [60]. The reasoning behind this double-ring arrangement brings us a more optical perspective on SRRs. The current loop in a single-ring SRR can be

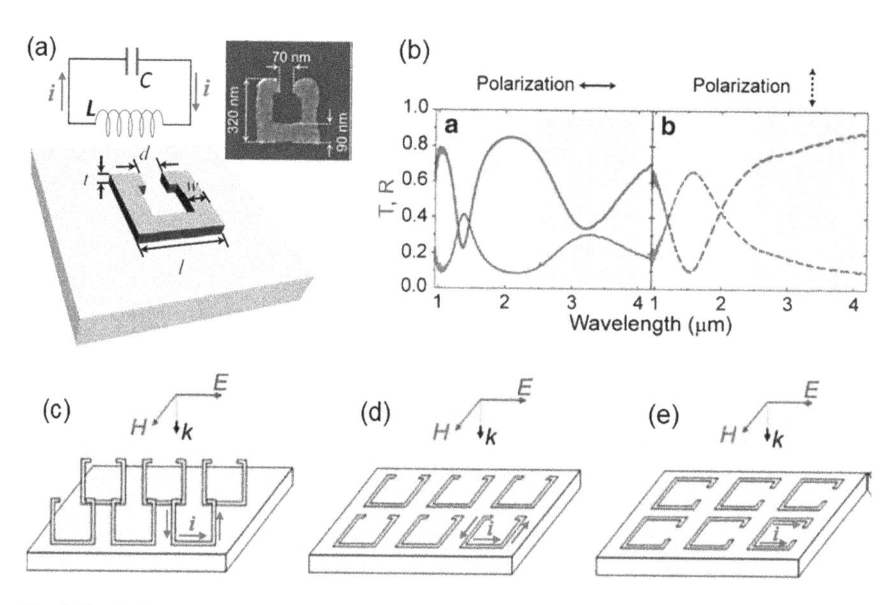

Fig. 2.12 Split-ring resonators (SRRs) as elements of artificial magnetism. (**a**) A single SRR on a substrate. Insets show the equivalent LC circuit (upper left) and SEM micrograph (upper right). (**b**) Transmission (red) and reflection (blue) spectra of an array of SRRs lying on a substrate. The incident field is polarized parallel (left) and perpendicular (right) to the gap of the SRRs, respectively. (**c**, **d**) Excitation of magnetic resonance in SRRs by the magnetic (**c**) or electric (**d**) component of the incident field. (**e**) No magnetic response can be excited when the magnetic field is parallel to the SRR plane and the electric field is perpendicular to the gap. Small red arrows denote the conduction current i. (Panels (a) and (b) are adapted from [51, 61] with permissions from Springer Nature and AAAS, respectively. Panels (c)–(e) are adapted from [51] with permission from Springer Nature)

treated as a magnetic dipole **m** perpendicular to the plane of SRR, and the oscillating charges near the gap act like an electric dipole **p** parallel to the gap. Therefore, a single-ring SRR at resonance exhibits bianisotropic coupling of electric and magnetic responses. Double-ring SRRs can effectively minimize this effect, because the electric dipole moments of the two rings largely cancel each other. Meanwhile, additional capacitance is introduced by the concentric rings. These advantages make the double-ring SRR a very successful design of magnetic elements and a prototype in the research of metamaterials at microwave frequencies. Nonetheless, for optical metamaterials, the preferred choice of building blocks is based on the single-ring structure.

The favor of single-ring SRRs over their double-ring cousins in the optical regime resides mainly on two facts. First, geometrical scaling is a basic strategy for shifting the operating wavelengths of SRRs. As suggested in [62–64], linear scaling is thought valid for frequencies below ~100 THz, where the kinetic energy of electrons in the metal can be neglected (as in an ideal metal). The resulting sizes of SRRs are in the range of several hundreds of nanometers, which is already challenging for fabrication of concentric structures. The inset of Fig. 2.12a shows the SEM micrograph of a single SRR with its magnetic resonance designed at ~3-μm wavelength [61]. The corresponding transmission and reflection spectra of a planar array consisting of such unit cells are presented in Fig. 2.12b. For frequencies above 100 THz, the linear scaling breaks down due to electron inertia, and for single-slit SRRs, the resonance frequency saturates near the boundary between visible and near-infrared light. Second, with minimized electric polarizability, double-ring SRRs can only be excited by aligning the magnetic field component of the incident light with the magnetic dipole moment of the SRRs. When SRRs are lying on a substrate, this requires oblique incidence. The excitation of single-ring SRRs, in contrast, has more options even at normal incident [51]. Figure 2.12c–e summarizes the possible configurations of alignment between SRRs on a substrate and the incident fields. The standard way of excitation, as illustrated in Fig. 2.12c, is having the magnetic field component of incident light penetrating the SRR plane, which will induce a current circulating in the ring. The problem with this scheme in practice is the difficulty of constructing vertical SRRs with micro- and nano-fabrication techniques. An alternative approach that works for SRRs lying on the substrate is shown in Fig. 2.12d. Because the magnetic field is parallel to the SRR plane, its interaction with the out-of-plane magnetic dipole moment is forbidden. The electric field component of the incidence, on the other hand, takes over the key role to induce the current loop. The critical factor of this excitation scheme is the orientation of the SRRs with respect to the polarization. When the electric field is polarized along the gap, currents are excited in both the gap-bearing arm and the one parallel to it but with different magnitudes. This asymmetric pair of current segments results in a net current loop in the SRR, thereby inducing an out-of-plane magnetic dipole. The other possible orientation has the electric field perpendicular to the gap (see Fig. 2.12e). In this case, currents are excited symmetrically in the two arms parallel to the polarization, which gives no magnetic response [65]. The different behaviors of SRRs under different polarizations can be clearly seen in Fig. 2.12b. The reflection peaks between 1- and 2-μm wavelengths correspond to the electric resonance of the metallic arms, which has a weak dependence on the polarization. In contrast, the magnetic resonance at ~3 μm is present only for the configuration in Fig. 2.12d.

The saturation of the magnetic response of SRRs at optical frequencies demands alternative designs of magnetic elements. In addition to introducing more slits to SRRs, a possible shortcut is taking advantage of the coupling between resonant electric elements. Two nanoparticles placed in proximity are strongly coupled. The interactions between two resonant plasmonic particles can cause new modes arising from hybridization of the plasmons supported by individual particles [67, 68]. Having in mind that the task is to find a structure where a current loop can be induced, transverse coupling of two nanorods turns out to be a sensible choice. Figure 2.13a plots the energy-level diagram of the plasmon hybridization for two identical nanorods in a transverse configuration. Two hybridized modes arise through the coupling between the dipolar resonances of the nanorods at frequency ω_0, of which the low-energy antisymmetric mode corresponds to a pair of antiparallel currents. Together with the displacement currents across the gaps near the ends of the nanorods (Fig. 2.13b), the antisymmetric currents constitute a magnetic dipole moment. An important note is that the nanorods need to be stacked in the direction of wave propagation; otherwise placing them side by side on a plane

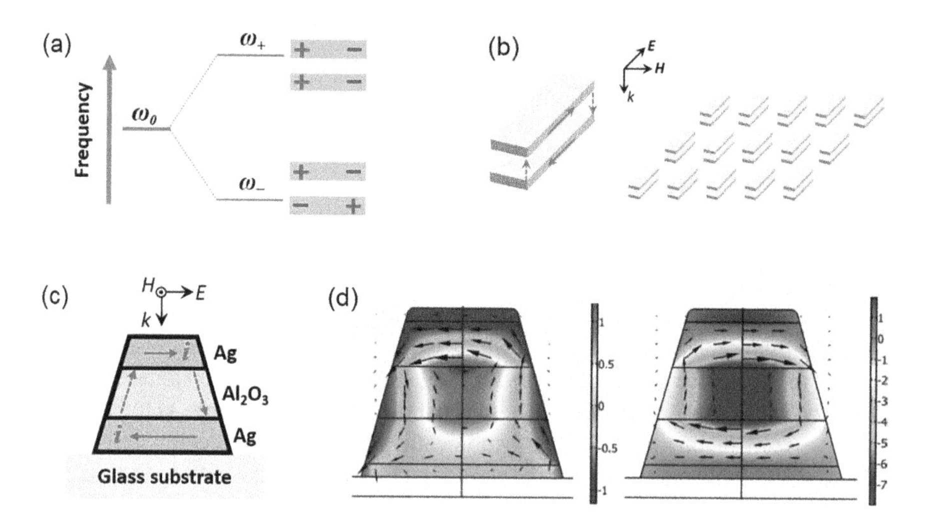

Fig. 2.13 (**a**) Energy-level diagram illustrating the transverse coupling of two identical dipolar resonators. The lower-energy antisymmetric mode is associated with antiparallel currents in the coupled resonators. (**b**) The antiparallel currents in paired nanorods and the displacement currents in the gaps form a current loop that gives rise to a magnetic dipole moment. Magnetic response arises in an array of these elementary cells. (**c**) Cross-sectional view of a current loop in a pair of trapezoidal-shaped nanostrips excited by TM-polarized light. (**d**) Simulated electric displacements (arrows) and magnetic field distributions (color map) of the electric (left) and magnetic (right) resonances in the structure in (**c**). (Panel (d) is reprinted from [66] with permission. © 2007 Optica Publishing Group)

perpendicular to the wavevector reduces the system's symmetry to the same situation as in Fig. 2.12e.

The idea of stacking paired resonators has been widely used for designing magnetic elements at optical frequencies. A representative design based on nanostrips is illustrated in Fig. 2.13c. The cross section of the structure highly resembles the stacked nanorods. Taking fabrication constraints into account, the sidewalls are purposely tapered so that the upper strip is slightly narrower [66]. The fabricated samples with varying widths and periodicities are demonstrated to exhibit magnetic responses across the entire visible wavelength range. And similar to the left panel in Fig. 2.12b, an electric resonance is also observed at shorter wavelengths. Figure 2.13d compares the field distributions of the electric and magnetic resonances. Concentrated magnetic fields in the spacer and circulating electric displacements, especially the antiparallel currents in the nanostrips, can be clearly seen for the magnetic resonance.

A very useful trick for creating optical magnetism is based on the concept of mirror image. Vertical "staples" with varying geometric parameters fabricated on a dielectric-coated gold substrate were demonstrated to exhibit negative permeability in the near- and mid-infrared regimes [73]. Although the working principle of this design has a clear picture related to the LC circuit, where the effective inductance and capacitance, respectively, come from the upper loop and gaps between the staple's footings and substrate, treating the structure as a staple and its image mirrored below the gold surface eases the analysis. An even simpler element can be obtained by sandwiching a metallic nanodisk and a metal substrate with a thin dielectric layer [74, 75]. At normal incidence, the nanodisk and its mirror image act just like the stacked nanorods or nanostrips discussed above.

We emphasize again that the elements of metamaterials must be much smaller than the wavelength of interest. These subwavelength building blocks are sometimes called "meta-atoms" to reflect their elementary role in constituting "metamaterials." One may wonder whether more abstract connections exist in this analogy. For instance, in chemistry, molecules with identical chemical formulae can have different spatial arrangements of atoms, known as isomers. If the properties of a metamaterial are attained from the elementary units, will the arrangement of them be a determining factor? The answer is, not surprisingly, yes. Global ordering or randomness aside, the alignment of elements within a unit cell largely determines the achievable responses, which can be much more complicated than negative permittivity/permeability resulting from resonant electric/magnetic dipole moments. Figure 2.14a depicts the concept of stereo-SRR dimer [69]. Two SRRs are stacked with different orientations, described by a twist angle. In this vertical coupling scheme, the hybridization of the modes supported by individual SRRs is sophisticated but can still be explained with the tools we have introduced so far [68]. For simplicity, let us discuss two special cases, where the twist angles are 0° and 90°, respectively. To ensure effective excitation, the field is incident from the top and polarized along the gap of the upper SRR. When the twist angle is 0°, the electric and magnetic dipolar resonances of both SRRs can be excited. The electric dipole moments along the gaps are coupled transversely as in Fig. 2.13a. Meanwhile, the

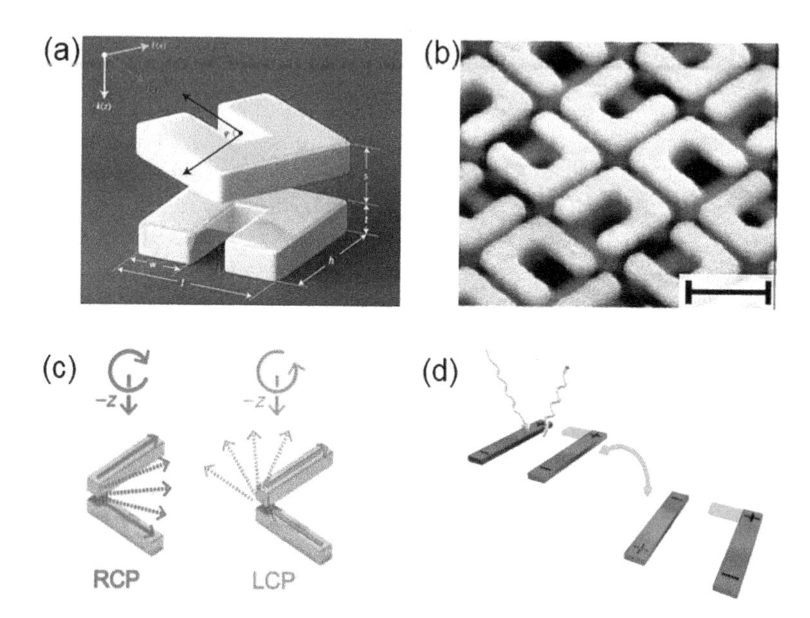

Fig. 2.14 (**a**) A stereo-SRR dimer containing stacked SRRs with a twist angle. (**b**) SEM micrograph of a two-layer chiral metamaterial. Scale bar: 400 nm. (**c**) Stacked and twisted gold nanorods as chiral building blocks and plasmonic realizations of the Born-Kuhn model. (**d**) An asymmetric pair of nanoantennas as chiral building blocks and photonic realizations of Fano resonances. Left: Bright mode at the higher energy level. Right: Dark mode at the lower energy level. (Panel (a) is reprinted from [69] with permission from Springer Nature. Panel (b) is reprinted from [70] with permission. © 2010 Optica Publishing Group. Panel (c) is reprinted from [71] with permission. Copyright (2013) American Chemical Society. Panel (d) is reprinted from [72] with permission from Springer Nature)

out-of-plane magnetic dipole moments interact in a longitudinal configuration, which, opposite to the transverse one, leads to parallel dipole moments at the lower resonance frequency and antiparallel moments at the higher frequency. The two processes of hybridization compete, and it was shown that in this case, the electric dipole-dipole interaction dominates. A similar analysis can be done for the twist angle of 180°. The situation becomes drastically different when the twist angle is 90°. Under the aforementioned excitation conditions, only the resonances of the upper SRR can be excited directly, and those of the lower SRR are accessed through coupling. Because of the orthogonal orientations of the SRRs, the electric dipoles are perpendicular to each other and have minimal interactions. Hence, the energy levels are determined by the longitudinal coupling of magnetic dipoles. As the twist angle varies continuously between these discrete states, the hybridized modes evolve in a complex manner, and contributions from higher-order electric multipoles are non-negligible for some intermediate positions. Therefore, the optical properties of the metamaterials composed of a lattice of such stereo-SRR dimers have a strong dependence on the twist angle.

The twisted SRRs have been used to construct chiral metamaterials. A structure is chiral if it is not superimposable on its mirror image [76–78]. In optics and photonics, the interest in chiral structures is derived from the fact that they interact with circularly polarized light (CPL) differently, resulting in many fascinating effects that enrich the research themes of chiral optics or simply chiroptics. The definition of chirality requires that a chiral metamaterial must possess no structural symmetry. In other words, no center of inversion or mirror planes are allowed. Helices are the most intuitive chiral structures [79], but the fabrication is generally more challenging. Stacking planar elements (which by themselves are achiral) with twists is thus regarded as an easy alternative. Figure 2.14b shows the SEM image of a chiral metamaterial comprised of stereo-SRR dimers with a twist angle of 90°. To avoid linear birefringence, four dimers are organized two-by-two in a super cell, rotated by 0°, 90°, 180°, and 270° with respect to the stacking axis [70]. Strong optical activity was demonstrated from this metamaterial. A simpler design based on corner-stacked and orthogonally twisted nanorods is sketched in Fig. 2.14c. The simplification of geometry, given that the fundamental resonance of a nanorod is the electric dipolar mode, allows the optical response of the nanorods pair to be viewed as two orthogonally coupled harmonic oscillators and interpreted by the Born-Kuhn model [71]. The vertical spacing between the rods determines the alignment of field vectors and the structure. Qualitatively, a quarter-wavelength spacing matches the twisting rate of the field vectors as a circularly polarized plane wave propagates, giving rise to hybridized bonding and antibonding modes that can be selectively excited by using CPL.

Due to mirror symmetry to the axis along the thickness direction, planar asymmetric elements are intrinsically achiral. However, this symmetry can be broken in the physical realizations by, e.g., introducing a substrate [80]. The resulting planar chiral structures are important building blocks for chiral metasurfaces. Figure 2.14d illustrates an example of paired asymmetric nanoantennas consisting of a nanorod and an L-shaped nanoparticle. This structure was first used in the study of Fano resonances in plasmonic metamaterials [72, 81–83]. The short arm of the L-shaped structure enables the coupling of super-radiant or "bright" electric dipolar mode (parallel currents in the left panel) to the sub-radiant or "dark" electric quadrupolar mode (antiparallel currents in the right panel), despite that the two nanoantennas are placed on the same plane perpendicular to the incident wavevector. Soon afterwards, the application of this geometry extends to chiral metasurfaces [84], because it exhibits different efficiencies for polarization conversions from left- to right-handed CPL and vice versa.

2.2.2 *Metamaterials*

In the preceding subsection, although not mentioned explicitly, we already base some discussions on the optical properties of metamaterials (e.g., the spectra of an SRR array in Fig. 2.12b), instead of those of the individual constituent elements, as the evidence of a desired resonance. Nonetheless, on many occasions, the function of a metamaterial is carried out by more than one resonance mode, and judicious arrangements of the elementary units (in the spatial, spectral, and temporal domains) are crucial. In this subsection, we briefly discuss the essential concepts of metamaterials, including how their responses can be expressed by effective and homogenized parameters and be realized by designing the structures based on elementary units.

A little preparation before going into details is an extension of the electromagnetic parameter space. The classification in Fig. 2.11a applies to isotropic materials, of which the permittivity and permeability are complex-valued scalars. These parameters become tensors if a material is anisotropic. In the following, we consider nonmagnetic materials with $\mu = 1$, and the tensor of permittivity has been diagonalized to the form

$$\hat{\varepsilon} = \begin{pmatrix} \varepsilon_{xx} & 0 & 0 \\ 0 & \varepsilon_{yy} & 0 \\ 0 & 0 & \varepsilon_{zz} \end{pmatrix}. \tag{2.12}$$

Here, x, y, and z denote the principal axes of the material. In naturally occurring materials, crystals exhibit optical anisotropy with all the diagonal elements being positive. A crystal is termed to be biaxial when ε_{xx}, ε_{yy}, and ε_{zz} all take different values and uniaxial when two of them are identical; the medium is isotropic otherwise. To see the difference anisotropy makes to the optical properties, it is helpful to first derive the dispersion relation. For a time-harmonic plane wave with angular frequency ω, the wave equation in an anisotropic medium is

$$\mathbf{k} \times (\mathbf{k} \times \mathbf{E}) + \omega^2 \mu_0 \varepsilon_0 \hat{\varepsilon} \mathbf{E} = 0, \tag{2.13}$$

where $\mathbf{k} = (k_x, k_y, k_z)$ is the wavevector. Equation (2.13) can be further rewritten to [85]

$$\begin{pmatrix} k_0^2 \varepsilon_{xx} - k_y^2 - k_z^2 & k_x k_y & k_x k_z \\ k_x k_y & k_0^2 \varepsilon_{yy} - k_x^2 - k_z^2 & k_y k_z \\ k_x k_z & k_y k_z & k_0^2 \varepsilon_{zz} - k_x^2 - k_y^2 \end{pmatrix} \begin{pmatrix} E_x \\ E_y \\ E_z \end{pmatrix} = 0, \tag{2.14}$$

where $k_0 = \omega \sqrt{\mu_0 \varepsilon_0}$ is the wavenumber in a vacuum. Solving Eq. (2.14) for biaxial media is possible, but we choose to continue with uniaxial materials, which are related to a class of metamaterials in the following discussion. Without loss of generality, we align the optic axis in the z direction so that $\varepsilon_{zz} = \varepsilon_{\parallel}$ and $\varepsilon_{xx} = \varepsilon_{yy} = \varepsilon_{\perp}$. Note that some literatures use the subscripts in the opposite way, where parallel and orthogonal are defined with respect to the plane perpendicular to the optic axis. After some algebra, the dispersion relations can be obtained from the nontrivial solution to Eq. (2.14):

$$\left(k_x^2 + k_y^2 + k_z^2 - \varepsilon_{\perp} k_0^2\right) \left(\frac{k_x^2 + k_y^2}{\varepsilon_{\parallel}} + \frac{k_z^2}{\varepsilon_{\perp}} - k_0^2\right) = 0. \tag{2.15}$$

The first term on the left side corresponds to a spherical isofrequency surface in the momentum space (or **k**-space), which describes the dispersion of waves polarized in the x-y plane, known as TE or ordinary waves. The latter naming comes from the fact that waves polarized perpendicular to the optic axis have a velocity independent of the propagation direction. The second term also defines a dispersion relation, but the associated waves are polarized in a plane containing the optic axis, known as TM or extraordinary waves. In a crystal with positive $\varepsilon_\perp$ and $\varepsilon_\parallel$, the isofrequency surfaces of extraordinary waves are ellipsoidal.

Unusual optical properties arise when anisotropy becomes extreme. Imagine an ellipsoidal isofrequency surface that is highly elongated. The wavevector along the major axis of the ellipsoid can thus be much greater than k_0, allowing sub-diffractional modes to propagate. Even more striking phenomena occur when one of the tensor components is negative. The isofrequency surfaces of extraordinary waves are no longer bounded but turned into hyperboloidal, giving rise to the concept of hyperbolic metamaterials [85–90]. Figure 2.15 depicts the isofrequency surfaces for two possible combinations of $\varepsilon_\perp$ and $\varepsilon_\parallel$, where the hyperboloids can be twofold (when $\varepsilon_\parallel < 0$, $\varepsilon_\perp > 0$) or onefold (when $\varepsilon_\perp < 0$, $\varepsilon_\parallel > 0$). According to a widely used classification, the former with one tensor component being negative is named type I, whereas the latter with two components being negative is type II. In some literatures, type I/II is alternatively called dielectric/metallic type [90], a reflection of their difference in optical responses. The unbounded nature of hyperbolic dispersions implies that in principle, waves with an infinitely large wavevector can be supported.

In nature, there exist some materials possessing hyperbolic dispersions in the optical regime. However, building hyperbolic metamaterials is still of great interest because of their simple realizations and flexibility of engineering anisotropy. By definition, regardless of the specific type, a hyperbolic metamaterial has metallic

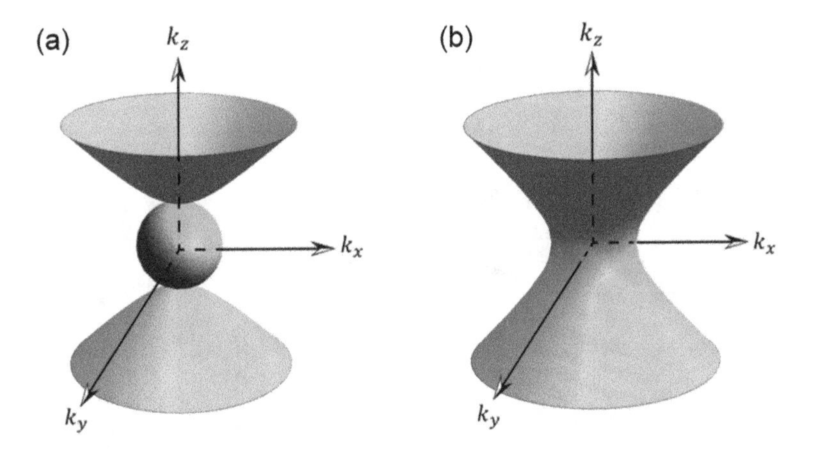

Fig. 2.15 Isofrequency surfaces of waves in hyperbolic metamaterials. (**a**) $\varepsilon_{zz} < 0$, $\varepsilon_{xx} = \varepsilon_{yy} > 0$. (**b**) $\varepsilon_{zz} > 0$, $\varepsilon_{xx} = \varepsilon_{yy} < 0$. The green sphere in (**a**) is the isofrequency surface of ordinary (TE-polarized) waves, and the blue hyperboloids in both panels are for extraordinary (TM-polarized) waves. (Reprinted from [90] with permission of AIP Publishing)

and dielectric behaviors along different principal axes. Therefore, the general strategy of its realization is to restrict the movement of free electrons within a metal along certain directions. Figure 2.16 illustrates two practical schemes, both of which manifest the design principle in a very straightforward fashion.

Next, let us prove the structures in Fig. 2.16 display hyperbolic dispersions. This works in the sense that the response of a metamaterial can be homogenized and expressed by effective material parameters [51, 91]. For the multilayer structure consisting of alternating metallic and dielectric thin films (Fig. 2.16a), the stack is not necessarily periodic, but all the layers are supposed to be deep subwavelength in thickness. We define the filling ratio of metal as $f = d_m/(d_m + d_d)$, where d_m and d_d are the total thicknesses of metallic and dielectric layers, respectively [87]. The effective permittivity tensor can be determined by considering the boundary conditions, which require the tangential components of the electric field across an interface to be continuous. In the system in study, this leads to

$$E_m^{\perp} = E_d^{\perp} = E_s^{\perp}, \tag{2.16}$$

where the superscripts indicate any direction parallel to the interface (i.e., perpendicular to the z-axis) and the subscripts m, d, and s denote the quantities' association with the metallic layer, dielectric layer, and the stack, respectively. The effective electric displacement in the tangential direction is taken as the averaged displacement in all layers:

$$D_s^{\perp} = fD_m^{\perp} + (1-f)D_d^{\perp}. \tag{2.17}$$

With the help of the constitutive relations, Eq. (2.17) can be rewritten as

$$\varepsilon_s^{\perp}E_s^{\perp} = f\varepsilon_m E_m^{\perp} + (1-f)\varepsilon_d E_d^{\perp}, \tag{2.18}$$

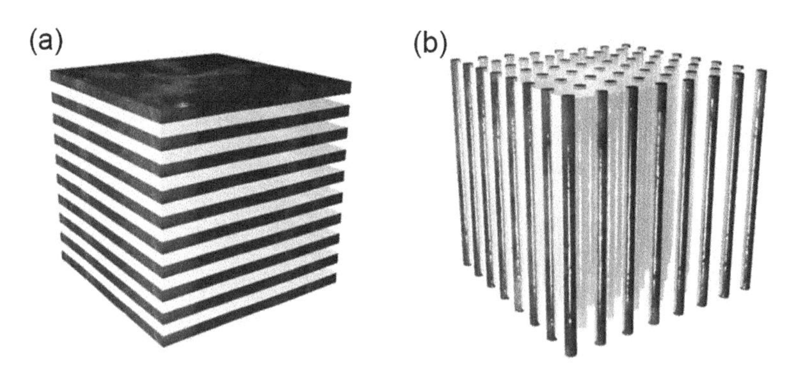

Fig. 2.16 Realizations of hyperbolic metamaterials. (**a**) Layered metal-dielectric structure. (**b**) Arrays of metallic nanowires embedded in a dielectric host. (Reprinted from [89] with permission. Copyright (2019) Wiley)

where the electric fields can be cancelled out as a result of Eq. (2.16). Therefore, we get the effective permittivity in the x-y plane:

$$\varepsilon_s^{\perp} = f\varepsilon_m + (1-f)\varepsilon_d. \tag{2.19}$$

Boundary conditions also require the normal components of the electric displacement across an interface to be continuous:

$$D_m^{\parallel} = D_d^{\parallel} = D_s^{\parallel}. \tag{2.20}$$

Similar to what has been done to the tangential electric displacement, by averaging the normal components of the electric field in all layers, the effective normal electric field is given by

$$E_s^{\parallel} = fE_m^{\parallel} + (1-f)E_d^{\parallel}, \tag{2.21}$$

and again, the electric fields can be cancelled out upon application of the constitutive relations, resulting in the final expression of effective permittivity in the z direction:

$$\frac{1}{\varepsilon_s^{\parallel}} = f\frac{1}{\varepsilon_m} + (1-f)\frac{1}{\varepsilon_d}, \tag{2.22a}$$

or

$$\varepsilon_s^{\parallel} = \frac{\varepsilon_m\varepsilon_d}{f\varepsilon_d + (1-f)\varepsilon_m}. \tag{2.22b}$$

For the metallic nanowire array in Fig. 2.16b, the tensor component in the direction parallel to the nanowires is attainable by applying the boundary conditions for tangential fields. The result has the same expression as Eq. (2.19), except that it applies to the direction parallel to the optic axis (z-axis):

$$\varepsilon_w^{\parallel} = f\varepsilon_m + (1-f)\varepsilon_d. \tag{2.23}$$

Here, the subscript w stands for wires, and the filling ratio f is defined based on the area occupation of metal in the x-y plane. Another method of describing the response of a periodic array of metallic wires to electric fields polarized along the wires is the effective Drude model [51]. The wire array has a diluted density of electrons and enhanced electron mass, resulting in significantly depressed effective plasma frequency. The derivation of the other tensor component is less straightforward than for multilayers. We present the expression below for completeness

$$\varepsilon_w^{\perp} = \frac{(1+f)\varepsilon_m\varepsilon_d + (1-f)\varepsilon_d^2}{(1+f)\varepsilon_d + (1-f)\varepsilon_m}, \tag{2.24}$$

whereas interested readers are referred to [87] for the details.

From Eqs. (2.19), (2.22b), (2.23), and (2.24), one can tell that the sign of each tensor component has strong dependence on the filling ratio and the permittivity of metal, which is highly dispersive in the optical regime. Based on noble metals and conventional dielectric materials, both designs in Fig. 2.16 can produce hyperbolic dispersions in certain wavelength ranges. Note that for a given structure, the dispersion relation it displays varies depending on the operating wavelength and can be type I or II hyperbolic or elliptical.

It is worth emphasizing again that the formulae derived above are the homogenized "effective" permittivity tensors of the metamaterials. They are valid in describing the macroscopic responses of the composite structures as a whole but not the microscopic interactions between the waves and each subwavelength constituent element. It is usually not possible to get analytical expressions when the building blocks have a complex architecture or are arranged in a random order. Numerous approaches have been developed to tackle this practical difficulty, including various forms of effective medium theory [91–93] and protocols [51] to retrieve the effective material parameters from measured or simulated transmission and reflection spectra.

The unbounded dispersion enabled by hyperbolic metamaterials has led to an important device for super-resolution imaging, termed as a hyperlens. Without reviewing the history of metamaterial research along this line [94–97], where the initial concept of perfect lens [98] evolves first to superlens [99] and then hyperlens [100], we focus on the role of hyperbolic metamaterials.

Light scattered off an object contains a spectrum of components in the momentum space. The components with low wavevectors are propagating waves that correspond to coarse spatial features, while the remaining components with high wavevectors are evanescent waves carrying the subwavelength information of the object. With conventional imaging devices, because evanescent waves cannot reach the far field, images reconstructed from propagating waves lack details and thus are diffraction-limited. The missing components with high wavevectors, however, are supported for propagation by structures with hyperbolic dispersion (Fig. 2.17a), allowing them to get involved in the image formation. Hence, when an object containing fine features is placed in proximity to a surface of a planar hyperbolic multilayer, super-resolution imaging can be achieved at the opposite surface.

The scheme can be modified for far-field super-resolution imaging, which requires conversion of evanescent waves to propagating waves. In addition to introducing designed scatterers at the output surface, an elegant method is constructing a hyperbolic metamaterial in a curved geometry. The most common design is based on concentric cylindrical or spherical multilayers [101]. The new solution brought in by this so-called hyperlens is a magnification mechanism. For example, in a concentric cylindrical multilayer (Fig. 2.17b), as waves propagate outwards, their tangential wavevectors are gradually compressed as a consequence of the conservation of angular momentum. Eventually, the wavevectors are small enough so that the associated waves become propagating in the surrounding medium, resulting in a magnified image in the far field. Multilayer structures with elliptical dispersion can

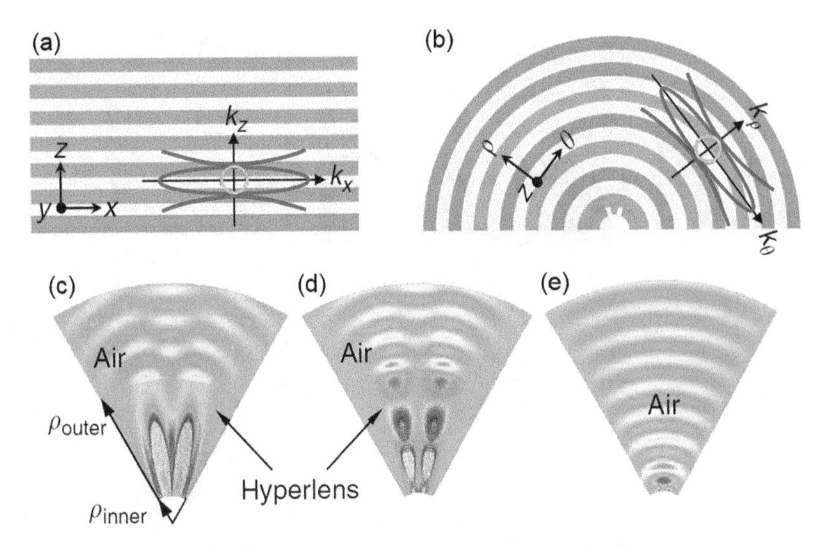

Fig. 2.17 (**a**) Planar multilayer metamaterials with hyperbolic (red curves) or elliptical (blue curve) dispersions. The orange circle represents the dispersion of light in an isotropic surrounding medium. (**b**) A hyperlens made of concentric multilayer metamaterials that display the same dispersions as in (**a**). (**c**–**e**) Far-field super-resolution imaging is attainable by using a hyperlens with elliptical (**c**) or hyperbolic (**d**) dispersion, but not by air alone (**e**). The objects are two line sources separated by a subwavelength distance of ~80 nm at the inner boundary. (Reprinted from [94] with permission from Springer Nature)

be used in the same way as along as the achievable wavevectors are large enough. Figure 2.17c, d shows the simulated field patterns for far-field super-resolution imaging using hyperlenses with elliptical and hyperbolic dispersions, respectively. In contrast, the same objects, i.e., two line sources with a deep-subwavelength separation, cannot be resolved in air without a hyperlens (Fig. 2.17e) [94].

Lastly, we discuss the design of metamaterials which exhibit a negative index of refraction. Located in the third quadrant in Fig. 2.11a, negative-index materials (NIMs) were once considered purely hypothetical, although many intriguing phenomena like negative refraction (Fig. 2.18a) had been predicted more than half a century ago [54]. The access to simultaneously negative values of ε and μ was opened by the invention and subsequent blooming of metamaterials [57], which were largely inspired by Sir John Pendry for his pioneering works on suggesting essential building blocks for NIMs and using NIMs as superlenses, among other new physics and applications [60, 98, 105–107]. Research efforts during the pursuit of NIMs were split to several streams, such as reexamining the phenomena associated with electromagnetic wave propagation and designing novel structures that operate at optical frequencies. Since the main purpose of this section is to help establish a basic understanding of the design principles of meta-atoms and metamaterials, here we focus on explaining two design examples of NIMs instead of any associated phenomena. Moreover, negative refraction can be realized on different platforms like anisotropic [108, 109] or chiral metamaterials [110]. In line with the

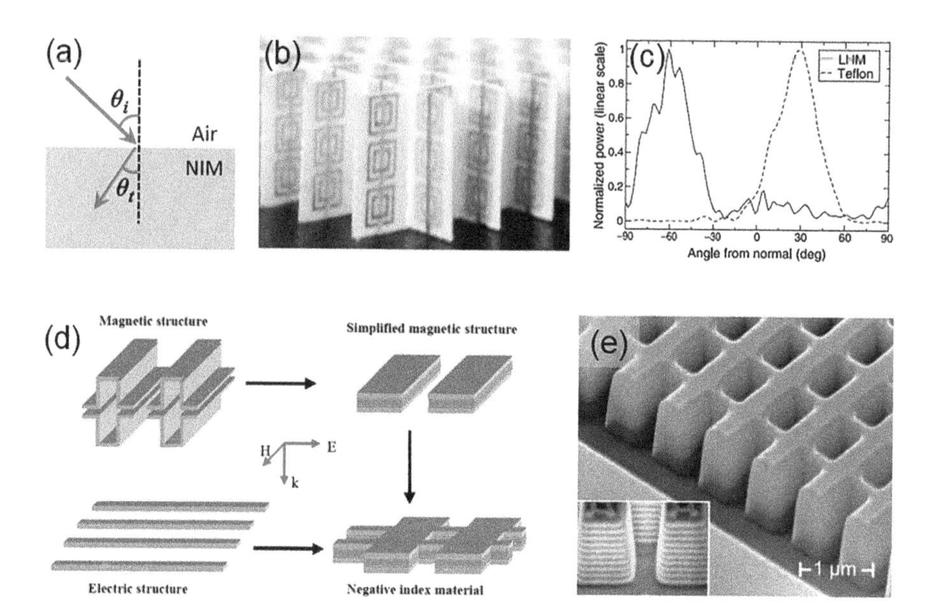

Fig. 2.18 (**a**) Scheme of negative refraction at the interface between air and a negative-index material (NIM). Following Snell's law, the refracted light beam is on the same side of the surface normal line as the incident light beam. (**b**) A microwave NIM consisting of periodic arrays of unit cells combining copper SRRs and wires on fiberglass circuit board material. (**c**) Experimental verification of negative refraction. (**d**) Design of NIM based on combined electric and magnetic resonant structures. (**e**) SEM image of fabricated fishnet structure. (Panels (b) and (c) are reprinted from [102] with permission of AAAS. Panel (d) is reprinted from [103] with permission. © 2005 Optica Publishing Group. Panel (e) is reprinted from [104] with permission from Springer Nature)

earlier preparation of building blocks, we restrict our discussion to the standard scheme based on simultaneously negative ε and μ.

The requirement of constructing an NIM is straightforward: overlapping two sets of elements with negative effective permittivity and permeability in the same frequency band. But this seemingly simple solution is not practical until the ingredient of artificial magnetism becomes available. Figure 2.18b shows the first experimental verification of a negative refractive index in microwave region [102]. The NIM consists of a 2D array of SRRs and metallic wires, both of which are lithographically deposited on opposite sides of circuit boards. The negative μ is provided by the SRRs at frequencies above their resonance, and the negative ε at matched frequencies is obtained by engineering the wire lattice so that its effective plasma frequency is brought down to microwave region but well above the SRR resonant frequency. When illuminated by a beam of microwaves with the electric field polarized along the wires, the structure exhibits simultaneously negative ε and μ. Negative refraction was demonstrated on a wedge-shaped sample, as shown in Fig. 2.18c.

Owing to the saturation of magnetic response of SRRs at high frequencies, construction of optical metamaterials with a negative refractive index must employ other building blocks, such as stacked nanorods or nanostrips. Interestingly, unlike in the microwave realization where two lattices of elements are interleaved, elementary structures with respective negative effective ε and μ at optical frequencies may be combined into one through judicious design. Figure 2.18d illustrates an example of this process [103]. Whereas the desired electric response is still from metallic wires, one can clearly see how the designs of magnetic meta-atoms evolve but manifest the same basic concept of a current loop, from SRRs, (mirror-imaged) vertical staple pairs, stacked nanostrips, and eventually the combined fishnet structures. The SEM image of a fabricated sample for near-infrared NIM is shown in Fig. 2.18e [104].

2.2.3 *Metasurfaces*

In the beginning of this section, we introduce metasurfaces as the 2D counterparts of metamaterials. It is also seen in the literature that metasurfaces are sometimes regarded as a class of metamaterials with a reduced dimensionality or, in other words, planar metamaterials [56]. The latter makes good sense in several aspects. For example, most metasurfaces are also made of arrays of subwavelength resonators, and they enable extraordinary optical responses not available from interfaces between two conventional materials. In addition, the significantly reduced losses and simplified fabrication of metasurfaces, thanks to their 2D nature, are in line with the pursuits in metamaterial research. Nonetheless, there are still enough reasons based on which we feel it is more appropriate to treat metasurfaces as a concept in an equal position to metamaterials. The most important one is the essence of metasurfaces in transforming waveforms is to create abrupt and spatially varying phase jumps with subwavelength thickness and resolution at an interface, which lifts the dependence of conventional optical devices on effects accumulated during light propagation in bulky materials, including metamaterials that are described by effective and homogenized parameters [5]. Since the milestone work published in 2011 by Capasso group [111], metasurfaces have become one of the major themes of nanophotonics research and revolutionized the landscape of flat optics, which has been covered by quite a few nice and timely reviews focusing on different topics [112–118]. Clarifications of the relation of metasurfaces to reflect-/transmit-arrays, frequency selective surfaces, and blazed gratings can be found in [5, 118–120] as well. Here, we again limit our discussion to a concise walk-through of the basics of metasurfaces for wavefront engineering. Working principles of the devices which perform other functions like novel absorption properties [121, 122], polarization conversion [123, 124], and manipulation of near fields [125, 126] or surface waves [127] can be deduced from these basics.

The significance of introducing phase shifts to an ordinary interface between two isotropic media can be seen by reexamining Snell's laws of reflection and refraction [111]. As illustrated in Fig. 2.19a, an interfacial phase gradient $d\Phi/dx$ in the plane of incidence optically acts as an effective wavevector in the same direction, which adds to the tangential component of the incident wavevector and in turn deflects the transmitted light as a consequence of wavevector conservation [119]. This change at the interface leads to the generalized law of refraction

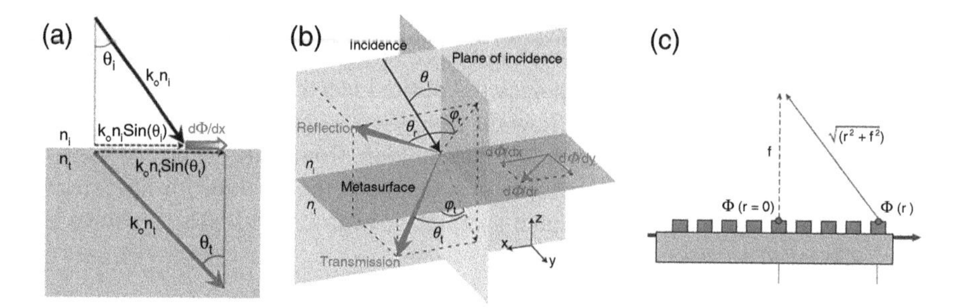

Fig. 2.19 (**a**) Schematic for deriving the generalized 2D Snell's law of refraction. The interface between the two media (with refractive indices n_i and n_t, respectively) hosts an array of optical antennas with subwavelength thickness and separation, which introduces a constant interfacial phrase gradient $d\Phi/dx$ in the plane of incidence, providing an effective wavevector along the interface that deflects the transmitted light beam. Although not shown, the same argument applies to reflection. Such anomalous responses can be as extreme as negative reflection/refraction or total conversion into surface waves. (**b**) In a more general 3D case, the phase gradient $d\Phi/d\mathbf{r}$ does not lie in the plane of incidence, further enabling out-of-plane reflection and refraction. (**c**) A metasurface with a properly designed phase profile Φ can carry out desired functions like lensing or forming structured light beams. Such control of light by flat optics can be multifunctional, multiwavelength, angle-/polarization-dependent, etc. (Panels (a) and (b) are reprinted from [119] with permission of IEEE. Panel (c) is reprinted from [114] with permission of Springer Nature)

$$n_t \sin\theta_t - n_i \sin\theta_i = \frac{\lambda_0}{2\pi}\frac{d\Phi}{dx}, \tag{2.25a}$$

where λ_0 is the wavelength in a vacuum. Meanwhile, the law of reflection is generalized to

$$\sin\theta_r - \sin\theta_i = \frac{\lambda_0}{2\pi n_i}\frac{d\Phi}{dx}. \tag{2.26a}$$

Equations (2.25a) and (2.26a) can be alternatively derived by using Fermat's principle. Similarly, when the phase gradient has a component perpendicular to the plane of incidence (along the y-axis in Fig. 2.19b), out-of-plane refraction and reflection occur [128], which are governed, respectively, by the following equations:

$$\cos\theta_t \sin\varphi_t = \frac{\lambda_0}{2\pi n_t}\frac{d\Phi}{dy}, \tag{2.25b}$$

$$\cos\theta_r \sin\varphi_r = \frac{\lambda_0}{2\pi n_i}\frac{d\Phi}{dy}. \tag{2.26b}$$

In theory, with a properly designed phase gradient, the refracted and reflected light can be directed into arbitrary directions.

Interfacial phase jumps are usually realized with optical scatterers (or antennas, if we take the extended definition) with subwavelength spacing and spatially varying geometric parameters. The requirement of subwavelength separation differentiates metasurfaces from diffractive optical components, and the spatial variation of scatterers carries out the phase profile needed to perform the desired optical functionality with a planar device. A good example to explain how the phase profile can be determined for a given functionality is the focusing of a normally incident plane wave, as illustrated in Fig. 2.19c. Because waves from different parts (described by the radial coordinate r) of an incident flat wavefront should remain in phase when arriving at the focal point, the flat lens needs to impart a phase profile $\Phi(r)$ that compensates the phase difference resulting from the different optical paths the refracted waves take, yielding

$$\Phi(r) = \frac{2\pi}{\lambda_0}\left(f - \sqrt{r^2 + f^2}\right), \tag{2.27}$$

where f is the focal length. The phase profiles of many other flat optical devices, such as flat axicons and lenses for generating structured light [119], non-diffracting beams [129], and hologram [130], can be derived following the same principle.

Thus far, everything we have been talking about is theoretical. The actual ability of controlling wavefront relies on the scattering properties of individual optical antennas. The complete control requires a full coverage of 2π phase range for the scattered light, along with sufficient amplitude and well-defined polarization. A single resonance of an optical antenna, nonetheless, only provides a tuning range of phase up to π [119]. Therefore, early demonstrations of metasurfaces utilize V-shaped plasmonic antennas [111, 132], which offer extended phase responses from two independent and orthogonal resonance modes, as shown in Fig. 2.20a. At a given wavelength, by choosing the arm lengths (h), spanning angles (Δ), and orientations of the antennas relative to the polarization of the incident light, the desired 0 to 2π phase coverage can be achieved for cross-polarized scattered light. Based on this library of two-oscillator scatterers, the first metasurface with a constant phase gradient was realized by periodically translating a super cell whose eight elements scatter mid-infrared light with nearly equal amplitudes and progressive phase delays of $\pi/4$ (see Fig. 2.20b, c), serving in the first demonstration of generalized laws of reflection and refraction.

A major limitation of metasurfaces using two-oscillator plasmonic antennas is the low efficiency, because the cross-polarized scattered light from a single layer of resonators carries only a small portion of the incident power. A simple yet very effective method to boost the efficiency, although it works for reflective metasurfaces only, is to place optical antennas at a small distance above a backplane as in reflect-arrays. The extended phase response is made possible for co-polarization even with simple scatterers like nanorods. The mechanism can be understood in two ways: (1) The antenna and its mirror image each support a dipolar resonance, equivalent to a two-oscillator element. (2) The antenna, spacer, and backplane form a

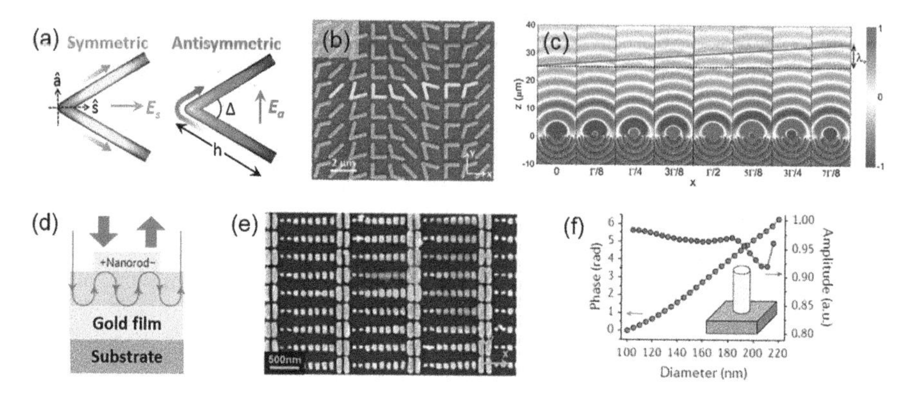

Fig. 2.20. (**a–c**) V-shaped plasmonic optical antennas as building blocks of metasurfaces. (**a**) A V-shaped optical antenna supports two orthogonal modes, which have symmetric (left) and anti-symmetric (right) current distributions and are excited by the incident field components along the $\hat{s}$ and $\hat{a}$ axes, respectively. (**b**) SEM image of a metasurface consisting of an array of V-shaped gold optical antennas on a silicon substrate, which generates a constant interfacial phase gradient along the x-axis. Highlighted is a super cell containing eight elements spanning a periodicity of $\Gamma = 11$ μm in the x-direction. (**c**) For a y-polarized plane wave excitation from the substrate ($z \leq 0$) at normal incidence, the cross-polarized scattered electric field (E_x) from individual antennas in the highlighted super cell in (**b**) exhibits a linear phase progression in steps of $\pi/4$, resulting in an anomalously refracted plane wave as indicated by the tilted wavefront (red line). (**d**) Fabry-Pérot cavity for reflect-array metasurfaces. (**e**) SEM image, with a super cell highlighted, of a metasurface to achieve high-efficiency anomalous reflection of y-polarized light. (**f**) Dielectric nanopillars are good choices for transmissive metalenses, with circular (anisotropic) cross sections suited for applications insensitive (sensitive) to polarization. (Panels (a)–(c) are reprinted from [111] with permission from AAAS. Panel (e) is reprinted from [131] with permission. Copyright (2012) American Chemical Society. Panel (f) is reprinted from [114] with permission of Springer Nature)

Fabry-Pérot cavity, where the reflected light excites the antenna repeatedly with a phase delay, as sketched in Fig. 2.20d. The reflective metasurfaces for demonstrating anomalous reflection (Fig. 2.20e) [131] and hologram [133] in the near-infrared region can reach efficiencies at high as ~80%, limited mainly by the absorption from the ohmic loss of metals. The general strategy for achieving high-efficiency transmission is to match the impedance of the metasurface with that of free-space Z_0. This requires rational design as well as optimization of the optical antennas. In the ideal case, if the surface electric and magnetic polarizabilities α_e and α_m, which can be related to equivalent surface electric and magnetic currents, are adjusted properly under the constraint [134]

$$\sqrt{\alpha_m / \alpha_e} = Z_0, \tag{2.28}$$

complete control over the transmitted light is attainable with no reflection. It is worth noting that under the above condition, individual antennas constituting the metasurface essentially function as Huygens sources. Therefore, this type of transmissive metasurfaces is sometimes referred to as Huygens surfaces.

One last class of building blocks widely used is high-refractive-index dielectric nanopillars. Although dielectric nanoparticles resonating under Kerker conditions can already serve as the elements of Huygens surfaces, nanopillars, with a wavelength-scale height and subwavelength lateral dimensions including spacing, provide another route to high transmission [135]. The working principle is not as easy to visualize as in the previous schemes. In contrast to those involving two resonances, the wavelength-scale height of the nanopillars allows for occurrence of high-order modes. In this regard, nanopillars can be considered as waveguides truncated on the top and bottom facets [136], which operate as resonators with low-quality factors. An attractive property of nanopillars is that at a given wavelength, the coverage of 2π phase range can be accomplished by varying their lateral dimensions with a fixed height, while the transmission remains high and nearly constant. Such correspondence between transmission, phase, and geometric parameters is easy to establish by simulations with simple parameter sweep. An example of circular nanopillars is shown in Fig. 2.20f.

In all the examples discussed above, the phase response comes from the dispersion of the antenna resonances, tuned by varying the shape of individual elements in the array. Another very important and fundamentally different approach to introducing interfacial phase jumps is to use the Pancharatnam-Berry (PB) phase [138, 139], which is a geometric phase realized by spatial-variant polarization manipulation. Following the earlier scheme proposed by Hasman and coworkers [140–142], the realization of PB phase in metasurfaces is based on anisotropic optical antennas (e.g., nanorods, nanofins, slit apertures, SRRs, etc.) that are identical in shape across the array but take spatially varying orientations [143–145]. The PB phase can be explained with an elegant picture on the Pioncaré sphere [112, 141]. However, the principle of its use in metasurfaces is more intelligible when put into the Jones calculus. Let us consider nanorods, for example. For such an anisotropic optical antenna, its Jones matrix is

$$M(\alpha) = R(-\alpha)\begin{pmatrix} t_l & 0 \\ 0 & t_s \end{pmatrix} R(\alpha). \tag{2.29}$$

Here, t_l and t_s are the forward scattering coefficients for the incident light polarized along the rod's long and short axes, respectively, α is the angle by which a nanorod is rotated relative to a chosen direction in the metasurface plane, and R is the rotation matrix

$$R(\alpha) = \begin{pmatrix} \cos\alpha & \sin\alpha \\ -\sin\alpha & \cos\alpha \end{pmatrix}. \tag{2.30}$$

When the incident light is circularly polarized, its electric field vector can be expressed by $(1 \pm i)^{\mathrm{T}}$, where the +/− sign corresponds to left−/right-handed CPL and T denotes the matrix transpose. The transmitted light is then given by [114].

$$\begin{pmatrix} E_x \\ E_y \end{pmatrix} = M(\alpha)\begin{pmatrix} 1 \\ \pm i \end{pmatrix} = \frac{t_l + t_s}{2}\begin{pmatrix} 1 \\ \pm i \end{pmatrix} + \frac{t_l - t_s}{2} e^{\pm i2\alpha}\begin{pmatrix} 1 \\ \mp i \end{pmatrix}. \tag{2.31}$$

Equation (2.31) reveals that the transmission consists of two CPL components. The first part is co-polarized with respect to the incidence and modulated by a constant coefficient, whereas the other one differs from it not only in handedness but by carrying an extra phase term $\exp(\pm i2\alpha)$ in the coefficient. Therefore, in the transmission, the cross-polarized CPL scattered from optical antennas taking different orientations (described by the rotation angle α) receives a relative phase $\pm 2\alpha$. Obviously, this geometric phase can sweep from 0 to 2π, and it is possible to optimize the antennas so that the amplitude of cross-polarized CPL is maximized (see Fig. 2.21a). The introduction of PB phase greatly simplifies the design and fabrication of metasurfaces. Wavefront engineering by various phase profiles, such as anomalous refraction (Fig. 2.21b) [137] and focusing (Fig. 2.21c) [145], was demonstrated with it soon after the advent of the metasurface prototype. Many state-of-the-art high-performance metalenses are also based on this approach.

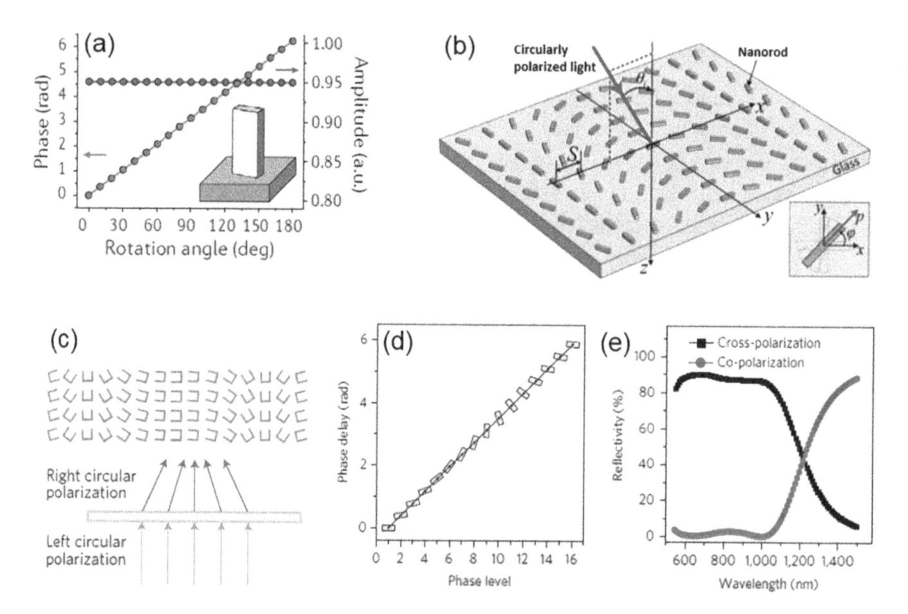

Fig. 2.21 Pancharatnam-Berry phase from identical anisotropic optical antennas with spatially varying orientations. (**a**) At a given wavelength, a dielectric nanofin can convert a large portion of circularly polarized incident light to its counterpart of opposite handedness (i.e., cross-polarized CPL) in transmission, with a phase shift twice the rotation angle of the nanofin. The co-polarized CPL component receives no such geometric phase. (**b**) Schematic of a metasurface consisting of a dipole antenna array. A super cell (in red) contains eight identical nanorods, separated by a distance S and rotated progressively with a step of $\pi/8$. The structure creates a constant phase gradient along the x-axis for transmitted CPL that is opposite in handedness to the incident one. (**c**) Schematic of split-ring apertures milled through a metal film used to focus CPL. (**d, e**) Geometric phase remains workable in the configuration of Fabry-Pérot cavity for reflected light, except that it applies to the CPL component with preserved handedness. (**d**) Phase delay as a function of gold nanorod orientation encoding phase level. (**e**) Reflectivity of co- and cross-polarized CPL under normal incidence. (Panel (a) is reprinted from [114] with permission from Springer Nature. Panel (b) is reprinted from [137] with permission. Copyright (2012) American Chemical Society. Panel (c) is reprinted from [5] with permissions of Springer Nature and Optica Publishing Group. Panels (d) and (e) are reprinted from [133] with permission of Springer Nature)

As a closing remark, the PB phase is applicable to reflected light as well. The only difference is, in reflection, it is associated with the co-polarized CPL component [133]. Figure 2.21d, e shows the phase chart and polarization-resolved reflectance for nanorods placed above a backplane (see Fig. 2.20d). High-efficiency and broadband metasurface holograms were demonstrated on this platform.

References

1. Bharadwaj, P., Deutsch, B., Novotny, L.: Optical antennas. Adv. Opt. Photon. **1**(3), 438–483 (2009)
2. Novotny, L., van Hulst, N.: Antennas for light. Nat. Photonics. **5**(2), 83–90 (2011)
3. Biagioni, P., Huang, J.-S., Hecht, B.: Nanoantennas for visible and infrared radiation. Rep. Prog. Phys. **75**(2), 024402 (2012)
4. Krasnok, A.E., et al.: Optical nanoantennas. Physics-Uspekhi. **56**(6), 539–564 (2013)
5. Yu, N., Capasso, F.: Flat optics with designer metasurfaces. Nat. Mater. **13**(2), 139–150 (2014)
6. Balanis, C.A.: Antenna Theory: Analysis and Design. Wiley, Hoboken (2015)
7. Novotny, L., Hecht, B.: Principles of Nano-optics, 2nd edn. Cambridge University Press, Cambridge (2012)
8. Girwidz, R.V.: Visualizing dipole radiation. Eur. J. Phys. **37**(6), 065206 (2016)
9. Purcell, E.M., Torrey, H.C., Pound, R.V.: Resonance absorption by nuclear magnetic moments in a solid. Phys. Rev. **69**(1–2), 37–38 (1946)
10. Pelton, M.: Modified spontaneous emission in nanophotonic structures. Nat. Photonics. **9**(7), 427–435 (2015)
11. Baranov, D.G., et al.: Modifying magnetic dipole spontaneous emission with nanophotonic structures. Laser Photonics Rev. **11**(3), 1600268 (2017)
12. Törmä, P., Barnes, W.L.: Strong coupling between surface plasmon polaritons and emitters: a review. Rep. Prog. Phys. **78**(1), 013901 (2014)
13. Ramezani, M., Berghuis, M., Gómez Rivas, J.: Strong light–matter coupling and exciton-polariton condensation in lattices of plasmonic nanoparticles [Invited]. J. Opt. Soc. Am. B. **36**(7), E88–E103 (2019)
14. Dovzhenko, D.S., et al.: Light–matter interaction in the strong coupling regime: configurations, conditions, and applications. Nanoscale. **10**(8), 3589–3605 (2018)
15. Caldarola, M., et al.: Non-plasmonic nanoantennas for surface enhanced spectroscopies with ultra-low heat conversion. Nat. Commun. **6**(1), 7915 (2015)
16. Rolly, B., et al.: Crucial role of the emitter–particle distance on the directivity of optical antennas. Opt. Lett. **36**(17), 3368–3370 (2011)
17. Bonod, N., et al.: Ultracompact and unidirectional metallic antennas. Phys. Rev. B. **82**(11), 115429 (2010)
18. Yao, K., Zheng, Y.: Directional light emission by electric and magnetic dipoles near a nanosphere: an analytical approach based on the generalized Mie theory. Opt. Lett. **46**(2), 302–305 (2021)
19. Hancu, I.M., et al.: Multipolar interference for directed light emission. Nano Lett. **14**(1), 166–171 (2014)
20. Liu, W., Kivshar, Y.S.: Generalized Kerker effects in nanophotonics and meta-optics. Opt. Express. **26**(10), 13085–13105 (2018)
21. Fu, Y.H., et al.: Directional visible light scattering by silicon nanoparticles. Nat. Commun. **4**(1), 1527 (2013)
22. Kerker, M., Wang, D.S., Giles, C.L.: Electromagnetic scattering by magnetic spheres. J. Opt. Soc. Am. **73**(6), 765–767 (1983)

23. Zambrana-Puyalto, X., et al.: Duality symmetry and Kerker conditions. Opt. Lett. **38**(11), 1857–1859 (2013)
24. Jin, P., Ziolkowski, R.W.: Metamaterial-inspired, electrically small Huygens sources. IEEE Antennas and Wirel. Propag. Lett. **9**, 501–505 (2010)
25. Geffrin, J.M., et al.: Magnetic and electric coherence in forward- and back-scattered electromagnetic waves by a single dielectric subwavelength sphere. Nat. Commun. **3**(1), 1171 (2012)
26. Evlyukhin, A.B., et al.: Demonstration of magnetic dipole resonances of dielectric nanospheres in the visible region. Nano Lett. **12**(7), 3749–3755 (2012)
27. Dorfmüller, J., et al.: Plasmonic nanowire antennas: experiment, simulation, and theory. Nano Lett. **10**(9), 3596–3603 (2010)
28. Novotny, L.: Effective wavelength scaling for optical antennas. Phys. Rev. Lett. **98**(26), 266802 (2007)
29. Dorfmüller, J., et al.: Fabry-Pérot resonances in one-dimensional plasmonic nanostructures. Nano Lett. **9**(6), 2372–2377 (2009)
30. Taminiau, T.H., Stefani, F.D., van Hulst, N.F.: Optical nanorod antennas modeled as cavities for dipolar emitters: evolution of sub- and super-radiant modes. Nano Lett. **11**(3), 1020–1024 (2011)
31. Ghenuche, P., et al.: Spectroscopic mode mapping of resonant plasmon nanoantennas. Phys. Rev. Lett. **101**(11), 116805 (2008)
32. Bidault, S., Mivelle, M., Bonod, N.: Dielectric nanoantennas to manipulate solid-state light emission. J. Appl. Phys. **126**(9), 094104 (2019)
33. Maksymov, I.S., et al.: Optical Yagi-Uda nanoantennas. Nanophotonics. **1**(1), 65–81 (2012)
34. Paniagua-Dominguez, R., Luk'yanchuk, B., Kuznetsov, A.I.: 3 – Control of scattering by isolated dielectric nanoantennas. In: Brener, I., et al. (eds.) Dielectric Metamaterials, pp. 73–108. Woodhead Publishing, Duxford (2020)
35. Curto, A.G., et al.: Unidirectional emission of a quantum dot coupled to a nanoantenna. Science. **329**(5994), 930–933 (2010)
36. Coenen, T., et al.: Directional emission from plasmonic Yagi–Uda antennas probed by angle-resolved cathodoluminescence spectroscopy. Nano Lett. **11**(9), 3779–3784 (2011)
37. Kosako, T., Kadoya, Y., Hofmann, H.F.: Directional control of light by a nano-optical Yagi–Uda antenna. Nat. Photonics. **4**(5), 312–315 (2010)
38. Kim, J., et al.: Babinet-inverted optical Yagi–Uda antenna for unidirectional radiation to free space. Nano Lett. **14**(6), 3072–3078 (2014)
39. Dregely, D., et al.: 3D optical Yagi–Uda nanoantenna array. Nat. Commun. **2**(1), 267 (2011)
40. Kullock, R., et al.: Electrically-driven Yagi-Uda antennas for light. Nat. Commun. **11**(1), 115 (2020)
41. Krasnok, A.E., et al.: All-dielectric optical nanoantennas. Opt. Express. **20**(18), 20599–20604 (2012)
42. Schmidt, M.K., et al.: Dielectric antennas – a suitable platform for controlling magnetic dipolar emission. Opt. Express. **20**(13), 13636–13650 (2012)
43. Mertens, H., Koenderink, A.F., Polman, A.: Plasmon-enhanced luminescence near noble-metal nanospheres: comparison of exact theory and an improved Gersten and Nitzan model. Phys. Rev. B. **76**(11), 115123 (2007)
44. Chigrin, D.N., et al.: Emission quenching of magnetic dipole transitions near a metal nanoparticle. ACS Photonics. **3**(1), 27–34 (2016)
45. Chew, H.: Transition rates of atoms near spherical surfaces. J. Chem. Phys. **87**(2), 1355–1360 (1987)
46. Klimov, V.V., Ducloy, M., Letokhov, V.S.: Spontaneous emission of an atom in the presence of nanobodies. Quantum Electron. **31**(7), 569–586 (2001)
47. Guzatov, D.V., Klimov, V.V., Poprukailo, N.S.: Spontaneous radiation of a chiral molecule located near a half-space of a bi-isotropic material. J. Exp. Theor. Phys. **116**(4), 531–540 (2013)

48. Klimov, V.V., et al.: Eigen oscillations of a chiral sphere and their influence on radiation of chiral molecules. Opt. Express. **22**(15), 18564–18578 (2014)
49. Alaeian, H., Dionne, J.A.: Controlling electric, magnetic, and chiral dipolar emission with PT-symmetric potentials. Phys. Rev. B. **91**(24), 245108 (2015)
50. Guzatov, D.V.: Radiative and nonradiative spontaneous decay rates for an electric quadrupole source in the vicinity of a spherical particle. J. Exp. Theor. Phys. **122**(4), 633–644 (2016)
51. Cai, W., Shalaev, V.M.: Optical Metamaterials. Springer, Cham (2010)
52. Ziolkowski, R.W.: Metamaterials: the early years in the USA. EPJ Appl. Metamater. **1**, 5 (2014)
53. Tretyakov, S.A.: A personal view on the origins and developments of the metamaterial concept. J. Opt. **19**(1), 013002 (2016)
54. Veselago, V.G.: Electrodynamics of substances with simultaneously negative Values of ε and μ. Usp. Fiz. Nauk. **92**(7), 517 (1967)
55. Jahani, S., Jacob, Z.: All-dielectric metamaterials. Nat. Nanotechnol. **11**(1), 23–36 (2016)
56. Kildishev, A.V., Boltasseva, A., Shalaev, V.M.: Planar photonics with metasurfaces. Science. **339**(6125), 1232009 (2013)
57. Smith, D.R., Pendry, J.B., Wiltshire, M.C.K.: Metamaterials and negative refractive index. Science. **305**(5685), 788–792 (2004)
58. Monticone, F., Alù, A.: The quest for optical magnetism: from split-ring resonators to plasmonic nanoparticles and nanoclusters. J. Mater. Chem. C. **2**(43), 9059–9072 (2014)
59. Hardy, W.N., Whitehead, L.A.: Split-ring resonator for use in magnetic resonance from 200–2000 MHz. Rev. Sci. Instrum. **52**(2), 213–216 (1981)
60. Pendry, J.B., et al.: Magnetism from conductors and enhanced nonlinear phenomena. IEEE Trans. Microwave Theory Tech. **47**(11), 2075–2084 (1999)
61. Linden, S., et al.: Magnetic response of metamaterials at 100 terahertz. Science. **306**(5700), 1351–1353 (2004)
62. Klein, M.W., et al.: Single-slit split-ring resonators at optical frequencies: limits of size scaling. Opt. Lett. **31**(9), 1259–1261 (2006)
63. Zhou, J., et al.: Saturation of the magnetic response of split-ring resonators at optical frequencies. Phys. Rev. Lett. **95**(22), 223902 (2005)
64. Tretyakov, S.: On geometrical scaling of split-ring and double-bar resonators at optical frequencies. Metamaterials. **1**(1), 40–43 (2007)
65. Mühlig, S., et al.: Multipole analysis of meta-atoms. Metamaterials. **5**(2), 64–73 (2011)
66. Cai, W., et al.: Metamagnetics with rainbow colors. Opt. Express. **15**(6), 3333–3341 (2007)
67. Prodan, E., et al.: A hybridization model for the plasmon response of complex nanostructures. Science. **302**(5644), 419–422 (2003)
68. Liu, N., Giessen, H.: Coupling effects in optical metamaterials. Angew. Chem. Int. Ed. **49**(51), 9838–9852 (2010)
69. Liu, N., et al.: Stereometamaterials. Nat. Photonics. **3**(3), 157–162 (2009)
70. Decker, M., et al.: Twisted split-ring-resonator photonic metamaterial with huge optical activity. Opt. Lett. **35**(10), 1593–1595 (2010)
71. Yin, X., et al.: Interpreting chiral nanophotonic spectra: the plasmonic Born–Kuhn model. Nano Lett. **13**(12), 6238–6243 (2013)
72. Wu, C., et al.: Fano-resonant asymmetric metamaterials for ultrasensitive spectroscopy and identification of molecular monolayers. Nat. Mater. **11**(1), 69–75 (2012)
73. Zhang, S., et al.: Midinfrared resonant magnetic nanostructures exhibiting a negative permeability. Phys. Rev. Lett. **94**(3), 037402 (2005)
74. Liu, Y., et al.: Compact magnetic antennas for directional excitation of surface plasmons. Nano Lett. **12**(9), 4853–4858 (2012)
75. Liu, N., et al.: Infrared perfect absorber and its application as plasmonic sensor. Nano Lett. **10**(7), 2342–2348 (2010)
76. Collins, J.T., et al.: Chirality and chiroptical effects in metal nanostructures: fundamentals and current trends. Adv. Opt. Mater. **5**(16), 1700182 (2017)

77. Hentschel, M., et al.: Chiral plasmonics. Sci. Adv. **3**(5), e1602735 (2017)
78. Mun, J., et al.: Electromagnetic chirality: from fundamentals to nontraditional chiroptical phenomena. Light Sci. Appl. **9**(1), 139 (2020)
79. Gansel, J.K., et al.: Gold helix photonic metamaterial as broadband circular polarizer. Science. **325**(5947), 1513–1515 (2009)
80. Menzel, C., Rockstuhl, C., Lederer, F.: Advanced Jones calculus for the classification of periodic metamaterials. Phys. Rev. A. **82**(5), 053811 (2010)
81. Miroshnichenko, A.E., Flach, S., Kivshar, Y.S.: Fano resonances in nanoscale structures. Rev. Mod. Phys. **82**(3), 2257–2298 (2010)
82. Luk'yanchuk, B., et al.: The Fano resonance in plasmonic nanostructures and metamaterials. Nat. Mater. **9**(9), 707–715 (2010)
83. Yao, K., Liu, Y.: Plasmonic metamaterials. Nanotechnol. Rev. **3**(2), 177–210 (2014)
84. Wu, C., et al.: Spectrally selective chiral silicon metasurfaces based on infrared Fano resonances. Nat. Commun. **5**(1), 3892 (2014)
85. Ferrari, L., et al.: Hyperbolic metamaterials and their applications. Prog. Quantum Electron. **40**, 1–40 (2015)
86. Poddubny, A., et al.: Hyperbolic metamaterials. Nat. Photonics. **7**(12), 948–957 (2013)
87. Shekhar, P., Atkinson, J., Jacob, Z.: Hyperbolic metamaterials: fundamentals and applications. Nano Convergence. **1**(1), 14 (2014)
88. Smolyaninov, I.I., Smolyaninova, V.N.: Hyperbolic metamaterials: novel physics and applications. Solid State Electron. **136**, 102–112 (2017)
89. Huo, P., et al.: Hyperbolic metamaterials and metasurfaces: fundamentals and applications. Adv. Opt. Mater. **7**(14), 1801616 (2019)
90. Guo, Z., Jiang, H., Chen, H.: Hyperbolic metamaterials: from dispersion manipulation to applications. J. Appl. Phys. **127**(7), 071101 (2020)
91. Choy, T.C.: Effective medium theory: principles and applications. Oxford University Press, Oxford (2015)
92. Sihvola, A.H.: Electromagnetic mixing formulas and applications. Institution of Electrical Engineers, London (1999)
93. Milton, G.W.: The theory of composites. Cambridge University Press, Cambridge (2002)
94. Lu, D., Liu, Z.: Hyperlenses and metalenses for far-field super-resolution imaging. Nat. Commun. **3**(1), 1205 (2012)
95. Zhang, X., Liu, Z.: Superlenses to overcome the diffraction limit. Nat. Mater. **7**(6), 435–441 (2008)
96. Kawata, S., Inouye, Y., Verma, P.: Plasmonics for near-field nano-imaging and superlensing. Nat. Photonics. **3**(7), 388–394 (2009)
97. Padilla, W.J., Averitt, R.D.: Imaging with metamaterials. Nat. Rev. Phys. **4**(2), 85–100 (2022)
98. Pendry, J.B.: Negative refraction makes a perfect lens. Phys. Rev. Lett. **85**(18), 3966–3969 (2000)
99. Fang, N., et al.: Sub-diffraction-limited optical imaging with a silver superlens. Science. **308**(5721), 534–537 (2005)
100. Liu, Z., et al.: Far-field optical hyperlens magnifying sub-diffraction-limited objects. Science. **315**(5819), 1686–1686 (2007)
101. Jacob, Z., Alekseyev, L.V., Narimanov, E.: Optical hyperlens: far-field imaging beyond the diffraction limit. Opt. Express. **14**(18), 8247–8256 (2006)
102. Shelby, R.A., Smith, D.R., Schultz, S.: Experimental verification of a negative index of refraction. Science. **292**(5514), 77–79 (2001)
103. Zhang, S., et al.: Near-infrared double negative metamaterials. Opt. Express. **13**(13), 4922–4930 (2005)
104. Valentine, J., et al.: Three-dimensional optical metamaterial with a negative refractive index. Nature. **455**(7211), 376–379 (2008)
105. Pendry, J.B., et al.: Transformation optics and subwavelength control of light. Science. **337**(6094), 549–552 (2012)

106. Pendry, J.B., et al.: Extremely low frequency plasmons in metallic mesostructures. Phys. Rev. Lett. **76**(25), 4773–4776 (1996)
107. Pendry, J.B., Schurig, D., Smith, D.R.: Controlling electromagnetic fields. Science. **312**(5781), 1780–1782 (2006)
108. Podolskiy, V.A., Narimanov, E.E.: Strongly anisotropic waveguide as a nonmagnetic left-handed system. Phys. Rev. B. **71**(20), 201101 (2005)
109. Yao, J., et al.: Optical negative refraction in bulk metamaterials of nanowires. Science. **321**(5891), 930–930 (2008)
110. Pendry, J.B.: A chiral route to negative refraction. Science. **306**(5700), 1353–1355 (2004)
111. Yu, N., et al.: Light propagation with phase discontinuities: generalized Laws of reflection and refraction. Science. **334**(6054), 333–337 (2011)
112. Chen, H.-T., Taylor, A.J., Yu, N.: A review of metasurfaces: physics and applications. Rep. Prog. Phys. **79**(7), 076401 (2016)
113. Shaltout, A.M., Shalaev, V.M., Brongersma, M.L.: Spatiotemporal light control with active metasurfaces. Science. **364**(6441), eaat3100 (2019)
114. Chen, W.T., Zhu, A.Y., Capasso, F.: Flat optics with dispersion-engineered metasurfaces. Nat. Rev. Mater. **5**(8), 604–620 (2020)
115. Dorrah, A.H., Capasso, F.: Tunable structured light with flat optics. Science. **376**(6591), eabi6860 (2022)
116. Lalanne, P., Chavel, P.: Metalenses at visible wavelengths: past, present, perspectives. Laser Photonics Rev. **11**(3), 1600295 (2017)
117. Li, G., Zhang, S., Zentgraf, T.: Nonlinear photonic metasurfaces. Nat. Rev. Mater. **2**(5), 17010 (2017)
118. Genevet, P., et al.: Recent advances in planar optics: from plasmonic to dielectric metasurfaces. Optica. **4**(1), 139–152 (2017)
119. Yu, N., et al.: Flat optics: controlling wavefronts with optical antenna metasurfaces. IEEE J. Sel. Top. Quantum Electron. **19**(3), 4700423–4700423 (2013)
120. Larouche, S., Smith, D.R.: Reconciliation of generalized refraction with diffraction theory. Opt. Lett. **37**(12), 2391–2393 (2012)
121. Ra'di, Y., Simovski, C.R., Tretyakov, S.A.: Thin perfect absorbers for electromagnetic waves: theory, design, and realizations. Phys. Rev. Appl. **3**(3), 037001 (2015)
122. Alaee, R., Albooyeh, M., Rockstuhl, C.: Theory of metasurface based perfect absorbers. J. Phys. D. Appl. Phys. **50**(50), 503002 (2017)
123. Grady, N.K., et al.: Terahertz metamaterials for linear polarization conversion and anomalous refraction. Science. **340**(6138), 1304–1307 (2013)
124. Yang, Y., et al.: Dielectric meta-reflectarray for broadband linear polarization conversion and optical vortex generation. Nano Lett. **14**(3), 1394–1399 (2014)
125. Yao, K., Zheng, Y.: Near-ultraviolet dielectric metasurfaces: from surface-enhanced circular dichroism spectroscopy to polarization-preserving mirrors. J. Phys. Chem. C. **123**(18), 11814–11822 (2019)
126. Wang, Z., et al.: Manipulating Smith-Purcell emission with babinet metasurfaces. Phys. Rev. Lett. **117**(15), 157401 (2016)
127. Sun, S., et al.: Gradient-index meta-surfaces as a bridge linking propagating waves and surface waves. Nat. Mater. **11**(5), 426–431 (2012)
128. Aieta, F., et al.: Out-of-plane reflection and refraction of light by anisotropic optical antenna metasurfaces with phase discontinuities. Nano Lett. **12**(3), 1702–1706 (2012)
129. Li, Z., et al.: Graphene plasmonic metasurfaces to steer infrared light. Sci. Rep. **5**(1), 12423 (2015)
130. Ni, X., Kildishev, A.V., Shalaev, V.M.: Metasurface holograms for visible light. Nat. Commun. **4**(1), 2807 (2013)
131. Sun, S., et al.: High-efficiency broadband anomalous reflection by gradient meta-surfaces. Nano Lett. **12**(12), 6223–6229 (2012)

132. Ni, X., et al.: Broadband light bending with plasmonic nanoantennas. Science. **335**(6067), 427–427 (2012)
133. Zheng, G., et al.: Metasurface holograms reaching 80% efficiency. Nat. Nanotechnol. **10**(4), 308–312 (2015)
134. Pfeiffer, C., Grbic, A.: Metamaterial Huygens' surfaces: tailoring wave fronts with reflectionless sheets. Phys. Rev. Lett. **110**(19), 197401 (2013)
135. Arbabi, A., et al.: Subwavelength-thick lenses with high numerical apertures and large efficiency based on high-contrast transmitarrays. Nat. Commun. **6**(1), 7069 (2015)
136. Arbabi, A., et al.: Dielectric metasurfaces for complete control of phase and polarization with subwavelength spatial resolution and high transmission. Nat. Nanotechnol. **10**(11), 937–943 (2015)
137. Huang, L., et al.: Dispersionless phase discontinuities for controlling light propagation. Nano Lett. **12**(11), 5750–5755 (2012)
138. Pancharatnam, S.: Generalized theory of interference, and its applications. Proc. Indian Acad. Sci. Sect. A. **44**(5), 247–262 (1956)
139. Berry, M.V.: Quantal phase factors accompanying adiabatic changes. Proc. R. Soc. Lond. A Math. Phys. Sci. **392**(1802), 45–57 (1984)
140. Bomzon, Z., et al.: Space-variant Pancharatnam–Berry phase optical elements with computer-generated subwavelength gratings. Opt. Lett. **27**(13), 1141–1143 (2002)
141. Bomzon, Z.E., Kleiner, V., Hasman, E.: Pancharatnam–Berry phase in space-variant polarization-state manipulations with subwavelength gratings. Opt. Lett. **26**(18), 1424–1426 (2001)
142. Hasman, E., et al.: Polarization dependent focusing lens by use of quantized Pancharatnam–Berry phase diffractive optics. Appl. Phys. Lett. **82**(3), 328–330 (2003)
143. Lin, D., et al.: Dielectric gradient metasurface optical elements. Science. **345**(6194), 298–302 (2014)
144. Khorasaninejad, M., et al.: Metalenses at visible wavelengths: diffraction-limited focusing and subwavelength resolution imaging. Science. **352**(6290), 1190–1194 (2016)
145. Kang, M., et al.: Wave front engineering from an array of thin aperture antennas. Opt. Express. **20**(14), 15882–15890 (2012)

Chapter 3
Fundamentals of Machine Learning

Abstract Machine learning (ML) is a subfield of broader artificial intelligence (AI) that through programming gives computers the ability to learn from data (Alpaydin E, Introduction to machine learning. MIT Press, Cambridge, 2014; Géron A, Hands-on machine learning with Scikit-learn, Keras, and TensorFlow: concepts, tools, and techniques to build intelligent systems. O'Reilly Media, Inc., Sebastapol, 2019; Goodfellow I, Bengio Y, Courville A, Deep learning. MIT Press, Cambridge, 2016). Owning advantages in tackling problems that are too complex for explicitly programmed algorithms, ML is considered as a powerful technique for making data-driven predictions and decisions. Early applications of ML, launched a few decades ago, were restricted to some specialized tasks, such as pattern recognition. With explosive growths of data and computational power over the succeeding decades, nowadays ML can be found in numerous applications even in our daily lives, from filtering email spam, ranking the search results, recommending videos and posts to language translation. The use of ML also extends widely into scientific domains, among which optics and photonics are, surprisingly or not, actively involved. In this chapter, we discuss the basics of ML in a descriptive manner. In particular, focus will be devoted to its most vibrant subfield of deep learning (DL) (Goodfellow I, Bengio Y, Courville A, Deep learning. MIT Press, Cambridge, 2016; Kelleher JD, Deep learning. MIT Press, Cambridge, 2019).

3.1 Introduction

3.1.1 Brief History of Machine Learning and Deep Learning

ML encompasses a vast variety of algorithms, which share in common the goal of learning patterns from data and generalizing them to make accurate predictions [1]. The current umbrella of modern ML techniques has developed over decades from sources in different fields. Tracking the complete history of ML or even the

The original version of this chapter was revised. The correction to this chapter is available at https://doi.org/10.1007/978-3-031-20473-9_7

K. Yao, Y. Zheng, *Nanophotonics and Machine Learning*, Springer Series in Optical Sciences 241, https://doi.org/10.1007/978-3-031-20473-9_3

narrowed class of DL is not only difficult but probably unnecessary for nonspecialists. Since the purpose of this chapter is to prepare the reader with essential knowledge for the later discussion on photonics-related applications, most of which center around DL, here we only summarize a few influential events in the development of DL. More comprehensive yet still compact surveys are available in many (text) books such as [3, 4] or online resources.

At the very core of DL is the concept of artificial neural networks (ANNs) or interchangeably neural networks (NNs). ANNs were inspired by the biological processes of neurons, with connections between nodes aimed at mirroring how synapses work together in the human brain. Attempts to model neural connections computationally as neural networks date back to the 1940s and 1950s (Fig. 3.1). In 1943, McCulloch and Pitts proposed a model, now regarded as the first ANN architecture (the neurons were in a rather primitive form though), to describe nervous activity with propositional logic [5]. The next milestone was the development of Hebbian theory, which described an important idea of how the repeatedly firing connections between cells (neurons) led to growth and learning over time [6]. The theory postulated that the efficiency of one synapse firing into another would increase with repeated attempts to fire, or "neurons that fire together wire together." Research on new models of neuron activity continued in the 1950s. In particular, the perceptron proposed and implemented by Rosenblatt enabled the training of a single neuron and served as an early prototype for image classification algorithms [7]. In 1971, Ivakhnenko published a paper on group method for data handling [8]. Overlooked for many years, this work provided the early blueprint for modern deep learning, enabling training of an eight-layer network. The neocognitron, developed in 1980, also showed the power of networks containing multiple hidden layers for visual pattern recognition [9]. These multilayer networks mirrored the concepts in Hebbian theory, with the connections between neurons in different layers being strengthened in the same way as synaptic connections. The procedure was further developed and made more feasible with backpropagation, an algorithm for efficiently calculating how to adjust the states of neurons in multilayer networks. Backpropagation was first implemented in conjunction with neural networks by Rumelhart, Hinton, and Williams in 1986 [10] and is probably the most important algorithm in DL. While there was progress on many fronts throughout the succeeding years, such as the invention of some important network architectures, the available computational power at the time made any large-scale practical implementations infeasible. After entering the new millennium, especially since the mid-2000s, the

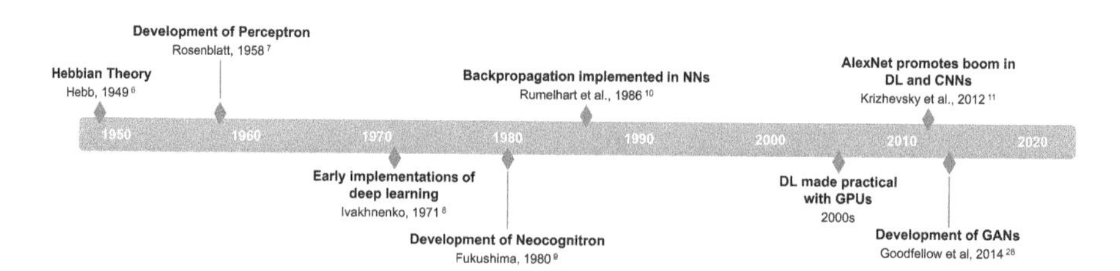

Fig. 3.1 Timeline of some milestones in the development of machine learning and deep learning

massive rise in computational resources through increased transistor counts and the development of graphics processing units (GPUs) helped boost the research of neural networks. It became possible to implement network structures with large numbers of layers and hundreds of thousands to millions of internal connections. These highly complex neural networks came to be described as "deep neural networks," and the associated subfield of ML utilizing them was termed as "deep learning." By the mid-2010s, deep neural networks have exploded in popularity, winning open contests and matching/surpassing human performance on certain recognition tasks and games as sophisticated as Go [11, 12]. And the wave is still ongoing.

Parallel to the development of neural networks, many other ML algorithms, such as random forests, *k*-nearest neighbors, linear/logistic regression, support vector machines, gradient boosting, and various clustering techniques, have been invented and continuously improved over the years. These algorithms also possess the ability to use data to develop a model's internal logic and capability to make predictions and/or gain knowledge. However, the goals of the model and the mechanism of training would differ considerably between different approaches. Each algorithm comes with pros and cons for different tasks and scenarios, depending on the goals of the model and the data available.

3.1.2 Categorization of Machine Learning Techniques

ML techniques can broadly be split into three categories: supervised learning, unsupervised learning, and reinforcement learning [1]. In some works, semi-supervised learning is considered as another one. These categories are defined by what type of data and how the data are utilized for training as well as the end goal of the trained model. Generally, the most basic out-of-the-box implementation of a certain algorithm will fit into one specific category. However, in tackling practical problems, most algorithms have numerous variations where some may fall into a different category or fulfill the criteria of multiple categories (Fig. 3.2).

In supervised learning, the goal is to learn a mapping from the input to some output provided by a supervisor. The outputs as desired solutions are called labels, and therefore, the datasets for supervised algorithms are labeled data consisting of examples of input-output pairs. The input is typically a vector, and the output can be one or more values depending on the tasks. The final goal of a model after training is to take in an input vector and produce an output prediction as close to the true output value, often called "ground truth," as possible. At this point, it will be helpful to make a clarification on two terms in ML, algorithms and models. In simple words, an algorithm is a procedure of how a problem can be solved, while a model is a computer program that implements a specific algorithm and processes data to solve the problem. The differences between these two terms in many contexts are somewhat subtle, and it is common that they are used interchangeably, as we will be doing in the later parts of this book.

Over the course of training, a model is fed with the labeled dataset and gradually adjusts its parameters based on the computation of a cost function, which indicates

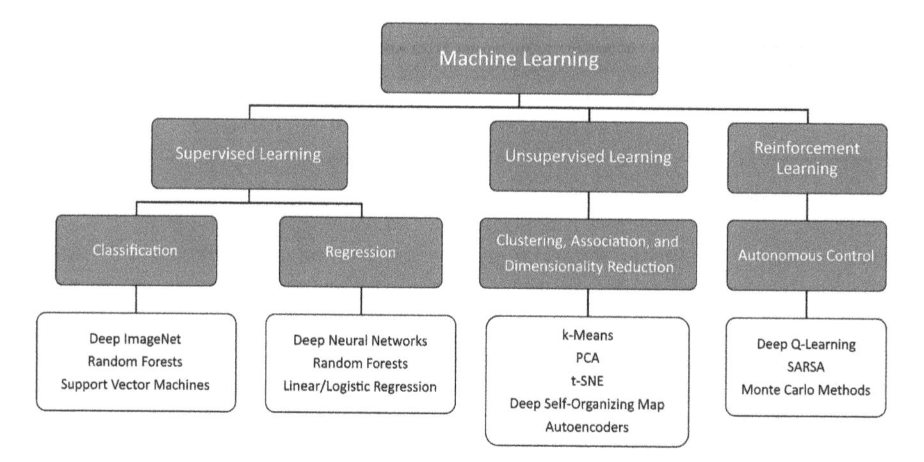

Fig. 3.2 Categorization of ML systems, typical tasks, and corresponding algorithms. Supervised and unsupervised learning build models off a training dataset, with the difference being whether the data include labels (desired solutions). Supervised learning can be further broken down into classification and regression tasks based on the output variable type. A certain class of algorithms, such as neural networks, can be used in different tasks with different implementations (e.g., deep self-organizing map, deep Q-learning, etc.). Although not shown in this diagram, semi-supervised learning is treated as a separate type of ML in some works. PCA, principal component analysis; t-SNE, t-distribution stochastic neighbor embedding; SARSA, state-action-reward-state-action

how well the model is making the prediction. As an example of supervised learning, a model could be trained to identify whether a cat is present in an image. For the training data, the input is the image represented by a matrix or vector of pixel values, and the output is the label given by a single binary number, 0 or 1, indicating whether a cat was present in that image. The model will be exposed to many images that either do or do not contain a cat and try to learn patterns that are indicative of whether the image contains a cat or not.

Typical supervised learning tasks can be split into two subcategories based on what the model is trying to predict: regression and classification. In regression tasks, the predicted output is one or more continuous variables that can take on any real-valued number. As a result, for each output variable, there are infinite possible values for the model to predict. In classification tasks, the predicted output is a categorical variable, meaning that it can take only one out of a finite number of possible values. These values can be viewed as "classes" or "categories" which the data can fall into, and the goal of the algorithms is to correctly classify the instance given the input variable. The previously listed example of cat identification is a classification task: it has two possible classes for the data, "contains cat" and "does not contain cat." A classification task can have more than two classes, e.g., sorting images of geometric shapes into predefined groups like circle, square, triangle, etc. Notably, however, the model cannot generalize to groups that are not included in the training dataset. If a model is trained only on recognizing circles, squares, and triangles, and an image of an octagon is fed in, the model will try to fit it into one of the three labeled categories. Therefore, it is important that the training data are representative of all the possible classes the model could expect to see and have a sufficient number of examples for each class to train a good classifier. Although a totally fair comparison would be difficult, regression is considered by many people to be easier than classification. The reason behind this belief is that classification is constrained to discrete values, and constrained problems are generally more difficult to solve. On some occasions, algorithms for classification are adapted to output continuous values, which will then be converted to class labels. For instance, if the label is binary like in the cat identification task, the model may output a real number in the interval [0,1], which gets converted to one state or the other based on a predefined threshold, like 0.5. Output values of 0.51 and 0.99 will both be interpreted as a prediction of the label "contains cat," while a larger value represents a higher confidence from the model in its prediction. Nonetheless, once converted to the same discrete value for a certain label, the outputs are equal.

Supervised algorithms have extraordinary predictive power, whereas the main drawback is the need for labeled data. Acquiring raw data can often be done quickly, but the requirement for labeled data implies that there needs to already have some other ways to make correct predictions. In many cases, this means manual labeling. Since many complex tasks require thousands to even several millions of training samples, it can be exceedingly slow and/or costly to acquire enough labeled data. Therefore, data acquisition is often the most significant bottleneck in ML.

Unsupervised learning is distinct from supervised learning in that the data does not have labels. In other words, the data are not paired and only consist of input vectors. Since there are no target values to predict, the goal of unsupervised learning is usually different from that of supervised learning. In unsupervised learning, the general course is to feed the model with raw data, letting it learn useful information or patterns in the input, while the user has less control over what the model is learning compared with supervised learning. One of the most common goals of unsupervised algorithms is clustering, i.e., splitting the instances in the datasets into distinct groups by some measure, mostly based on similarities or regularities. Taking text sorting as an example, an unsupervised learning algorithm can split papers and books into groups based on the occurrence frequency of certain words. This tends to coincide well with human-recognizable topics and genres, since those usually feature frequent use of some particular words and terms less commonly found elsewhere. To this end, unsupervised models can potentially perform the same task as supervised learning algorithms do, but notably, the same classification result is not guaranteed—it depends on the nature of the input data and the design of the cost function. Let us still consider the task of sorting geometric shapes. One can apply a clustering algorithm to a dataset of images of circles, squares, and triangles as before, but unlike with the supervised case, the clustering algorithm does not know those three categories while training. Therefore, it is possible that the algorithm ends up with some way other than the geometric shape to divide the images into several (not necessarily three) groups, if that criterion of categorization proves to be useful in increasing the "distance" between clusters. A core advantage of unsupervised learning is the algorithms' ability to find highly unintuitive structures or implicit patterns within the data, although this can also be a downside in the sense that such information is sometimes not interpretable.

Reinforcement learning (RL) is a third type of learning wholly separate from supervised and unsupervised ones. While the latter two are defined based on whether the training data are labeled or not, RL is an entirely different paradigm more closely linked to the conventional notions of soft AI, where the system, called an agent, attempts to act intelligently based on some reward measure. In RL, a set of scenarios (environment) needs to be defined in the first place, with actions an agent can take and observables it can measure at all times, including some assessments of its performance. The agent makes observations in the environment, takes actions, and gets rewards in return. The aim of the agent is to learn an optimal strategy (i.e., a sequence of correct actions), called a policy, to maximize the reward over time. Self-driving cars are a good example of a system that utilizes RL. In this case, the agent is the car, which can observe things about its environment, such as its own position and the positions of nearby objects (cars, pedestrians, road signs, etc.), and take any action like human drivers. A more detailed description of RL is given in the later section.

3.1.3 Popular Platforms to Train and Deploy Machine Learning Models

At some point of reading, interested readers may want to build a model or download one to see how things actually work. The core algorithms and techniques described later in this chapter can be implemented in many programming languages and integrated with commercial software, although some are more common and better supported for ML tasks than others. Python, an open-source and general-purpose programming language, is the most popular choice for ML modeling. It has numerous free and easily accessible libraries specialized for data engineering, fast matrix operations, and implemented ML algorithms. For data engineering purposes, for example, the libraries NumPy and pandas offer easy handling of large collections of data as well as fast matrix operations.

The most comprehensive library of ML algorithms is scikit-learn [13], usually abbreviated as sklearn. Sklearn features customizable implementations of a wide variety of over 100 popular ML models, termed as estimators, covering both supervised and unsupervised learning models with instructive examples. It also provides easy-to-use toolboxes for data preprocessing and model selection/evaluation. The main limitations of sklearn within the scope of this chapter are the relatively naive implementation of neural networks, mostly multilayer perceptron, and insufficient scalability, especially the lack of GPU support.

For neural networks, the most popular platform might be the open-source library TensorFlow [14]. TensorFlow was developed by Google as the second generation of an internal product for ML and DL research, later released to the public and is still actively updated. As the name suggests, it organizes data (usually in the form of multidimensional arrays) as tensors and is particularly strong in performing massive amounts of matrix operations required by the training and inference of deep neural networks. TensorFlow is designed to be run efficiently on both central processing units (CPUs) and GPUs, giving high flexibility for implementing ML on most hardware, from personal computers, mobile devices to high-performance computer clusters. Despite its extraordinary capability, TensorFlow is not the easiest to deploy. As a result, other libraries have been developed to make the creation of neural network models more user-friendly. Two of the most popular alternatives are Keras [15] and PyTorch [16]. Both libraries allow for quick and simple implementations of common network models, while the former has become the official high-level application programming interface of TensorFlow since the release of TensorFlow 2.0 in 2019.

3.2 Deep Neural Networks

3.2.1 *Basics of Artificial Neural Networks*

Let us now narrow down the discussion to artificial neural networks (ANNs). ANNs are composed of a large collection of processing units called artificial neurons, or simply neurons. These neurons, each containing a numerical value, are connected to other neurons in an interconnected graph, similar to biological neurons in the human brain. Just as brain cells send electrical signals to one another via synapses, neurons in an ANN will signal the other neurons they are connected to in the form of numerical transformations. For each connection between two neurons, there is a numerical value associated with it, called weight. Weights are parameters of a network to be adjusted during the training process. The functionality of neural interconnects, as illustrated in Fig. 3.3a, is to take the value of the previous neuron, modify that value with the associated weight by multiplication, and send the product to the output neuron. In the typical ANN architecture, a given neuron has many neurons connected to it and sending it signals, and the value of that neuron is calculated by two mathematical operations. The first one is the summation of all the values fed into the neuron via preceding connections. This essentially gives the weighted sum of all the values of the previous neurons. A trainable bias term is often added to the sum as well, introducing another degree of freedom to shift the linear function. The second operation is a nonlinear mapping, which further transforms the weighted sum and defines the value of the current neuron. The nonlinear mapping mimics the activation of a biological neuron, which fires when the received stimulations exceed a certain threshold. Therefore, the nonlinear mapping is also termed as the activation function. The nonlinear nature of activation functions is critical for an ANN to tackle complex representations in the data; otherwise linear mapping is inadequate on most occasions.

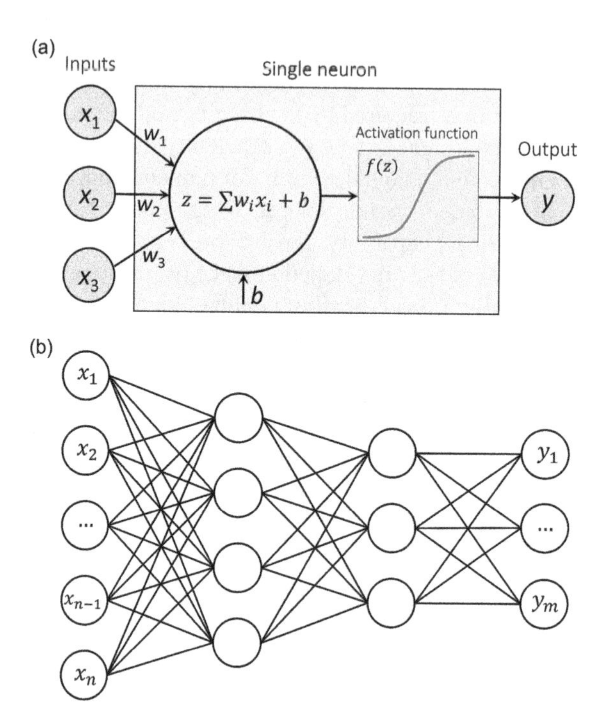

Fig. 3.3 Neurons and neural networks. (**a**) Connections of neurons into a neuron in the subsequent layer. The value of the neuron under consideration, denoted by y, is calculated by (1) taking the sum of the values of all the previous neurons (x_i), weighted by the strength of each corresponding connection w_i, and adding a bias term b to the weighted sum and (2) applying an activation function f to the summation. (**b**) For a deep fully connected neural network, this process is extended as every neuron from one layer is connected to every neuron in the adjacent layers. Values from the input neurons $[x_1, \ldots, x_n]$ are transformed multiple times before the network produces the values of the output neurons $[y_1, \ldots, y_m]$

In a typical neural network, neurons are organized into a series of layers, including an input layer that takes in the data and an output layer that gives the final numerical outputs (Fig. 3.3b). In between these two layers are one or more "hidden layers," with the naming reflecting the fact that these intermediate layers cannot be observed directly from a network's input or output. The number of hidden layers, or the depth of the network, is an important factor influencing an ANN's performance and level of training difficulty. ANNs with at least two, but usually many more, hidden layers are called deep neural networks. As already mentioned earlier, the associated subfield of ML that utilizes deep neural networks for modeling is termed as deep learning. In a deep neural network, the hidden layers close to the input/output layer are sometimes referred to as lower/upper layers.

The values in the hidden layers serve as intermediate states to aid in transforming the input data into the output predictions. These intermediate values often do not have easily interpretable meaning in relation to the data, and as a result, the internal functions of most ANNs are more akin to a black box. The general working mechanism of a neural network is as follows: Input data come into the first layer, and a series of numerical transformations are performed as the data pass through the hidden layers before arriving at the output layer. The transformations are determined by the weights of the connections between the neurons of successive layers, the bias terms, as well as the activation function at each neuron. The goal of training an ANN, as we will discuss next, is to optimize the learnable parameters, i.e., weights and biases, following some procedures such that the output prediction can be sufficiently accurate.

The training of neural networks includes two processes, forward mapping, where the input data are transformed through the hidden layers into an output prediction, and backpropagation, where the prediction from forward mapping is evaluated by a loss function in comparison to the true paired output (ground truth), and all the weights of the network are adjusted in the hope of making future predictions more accurate. These two processes are repeated continually, iterating through all the training data and adjusting the weights until convergence of the loss is reached, which indicates the network performance stops improving.

Now, let us walk through an example for a standard fully connected feedforward neural network to showcase the underlying mathematics. For the forward mapping, a single sample from the training dataset is fed in the form of an n-dimensional vector into the input layer of the network, resulting in a prediction in the form of an m-dimension vector at the output layer. Given the numerical values of the input, the values of neurons in the first hidden layer and in turn all the successive layers are calculated (Fig. 3.3b). The mathematical operations at each neuron, as described in the beginning of this subsection, can be expressed in the following form: for the i-th neuron in the n-th hidden layer, the summation is given by

$$z_i^n = \sum_{j=1}^{j=k} y_j^{n-1} w_{ij}^n + b_i^n, \tag{3.1}$$

where y_j^{n-1} is the value of the j-th neuron in the previous (input) layer, w_{ij}^n is the weight currently stored for the connection between y_j^{n-1} and the neuron in consideration, and b_i^n is the bias term. The resulting value of z_i^n, i.e., the weighted sum of values from all the neurons in the previous layer plus the bias, is then transformed by an activation function f to calculate the final value of the current neuron:

$$y_i^n = f\left(z_i^n\right). \tag{3.2}$$

Once the computation for the n-th layer is completed, the process is repeated for each successive layer until the final (say the N-th) output layer. Next, this predicted final output is compared to the ground truth output (also an m-dimensional vector t) paired with the input that is originally fed in the network. The comparison is performed by evaluating a loss function L, e.g., the mean square error (MSE)

$$L = \frac{1}{m}\sum_{i=1}^{m}\left(y_i^N - t_i\right)^2, \tag{3.3}$$

to see how much the network is “off” from the desired performance.

Evaluation of loss initiates the process of backpropagation, which is illustrated in Fig. 3.4. Backpropagation calculates the contribution of the individual weights to the loss and adjusts them accordingly to minimize the network's error. Gradient descent is the most common approach for this operation, by which the gradient of the loss with respect to every single weight is calculated, representing the contribution to the loss by each weight parameter. Since the end result of the loss is a nested function of various components that are each differentiable, the computation can be accomplished with simple chain rule of differentiation. This is simplest to see with the weights leading into the final output layer,

$$\frac{\partial L}{\partial w_{ij}^{N}} = \frac{\partial z_i^N}{\partial w_{ij}^N}\frac{\partial y_i^N}{\partial z_i^N}\frac{\partial L}{\partial y_i^N}. \tag{3.4}$$

When calculating the gradients with respect to the weights in the previous layer, the terms can be expanded to account for each value y_i^n being based on the values from the previous layer, as shown in Fig. 3.5 for the layer $N - 1$. As one goes back further and further in deeper networks, this could involve calculating millions of derivates for the update of all trainable parameters. The gradient represents the direction of steepest ascent for the loss function, and consequently, the negative of the gradient gives the direction of steepest descent, where the change in the weights would produce the biggest decrease in the loss function. After the gradients with respect to all weights are calculated, the weights are adjusted in the opposite direction of the gradient by taking a step

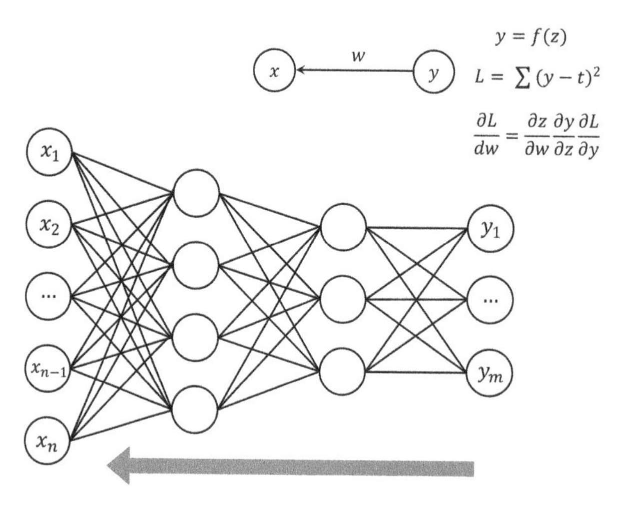

Fig. 3.4 Illustration of backpropagation. For a single neuron connection, the gradient of loss with respect to the weight associated with it can be calculated using the chain rule of derivatives. For a network with multiple layers, derivatives are expanded for all the connections moving backwards through the network

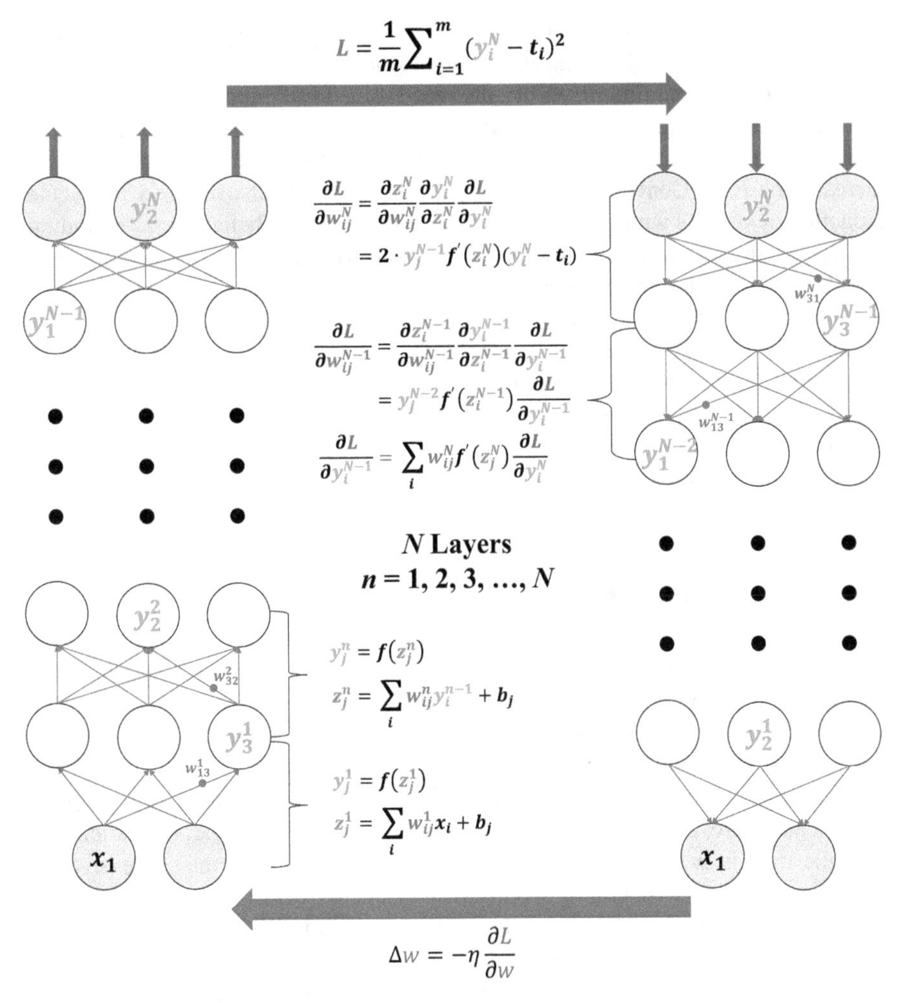

Fig. 3.5 Illustration of the procedures of forward mapping (left column) and backpropagation (right column) for a neural network with N layers in total. Neurons in the input and output layers are shaded. Input variables $\boldsymbol{x}$, neuron values $\boldsymbol{y}$, connection weights $\boldsymbol{w}$, weighted sums (with biases) z, and the loss function $\boldsymbol{L}$ are color-coded for clarity

$$\Delta w_{ij}^n = -\eta \frac{\partial L}{\partial w_{ij}^n}, \tag{3.5}$$

where the factor multiplied to the gradient term is called the learning rate η. The goal of the training process is to update the weights over and over again until convergence of the loss.

The learning process of an ANN is essentially an optimization; the possible values for all the weights form a vast space with some global minima of the loss computation. However, the loss is usually non-convex, meaning that if we optimize the loss with respect to the connection weights through gradient descent, instead of reaching the global minimum, we usually reach some local minima or saddle points. On the other hand, the negative of gradient represents the steepest descent at the current weight configuration, but moving too far in that direction could actually increase the loss. Conversely, moving only slightly will also mean a miniscule reduction in loss. As with other optimization problems, the challenge lies in finding the right middle ground to help the network converge to as low of a minimum as possible, with not too many steps so the optimization is done in a reasonable time. In general, it is prudent to have a relatively large learning rate at the beginning of training and decrease it over time, while the choice of the initial learning rate as well as how much and when to reduce it can make a big difference in the efficacy and speed of the training. For the most popular software implementations of ANNs, different types of optimizers exist to automate this process effectively. Nonetheless, some fine-tuning and trial and error will still be needed on a case by case basis.

3.2.2 *Activation Functions*

As mentioned before, the activation function is a mapping that is applied to the weighted sum of the inputs from the previous layer combined with a bias term. The activation function serves several purposes, the most important one of which is to add nonlinearity to the transformations. If simply taking the weighted sum of the neuron values with a bias term, we end up with a linear transformation of the inputs. And it can be proved that if every transformation in a stack is linear, no matter how deep the stack is or how complex the transformations are nested, the entire network is still linear [3]. The neural network has the same expressive power as a simple linear regression model while being far less efficient in learning. In order to model more complex representations, nonlinear transformations are necessary in the intermediate layers. Even with a small degree of nonlinearity at each neuron, when applied over the large scale of connections and nested over many layers, the network gains a massive expressive potential. Although any nonlinear function can add this extra expressive power in principle, a good choice of activation function for neural networks should meet at least a few conditions: (1) the function is differentiable everywhere; and (2) since the training process for a sufficiently complex ANN involves calculating the derivative many times, this calculation needs to be computationally inexpensive. Moreover, it should also avoid outputting values in the range of (0, 1) too often, as this can cause gradients to approach zero when multiplied over many successive layers, known as the vanishing gradient problem. These requirements on mathematical properties are general and do not decisively determine a network's performance. There are a number of commonly used activation functions in the training of ANNs, which suit different problems and tasks. In the following, we introduce a few popular ones. Another useful note is that in practical circumstances, it is possible that an ANN employs different types of activation functions for neurons in certain layers or modules.

The most common activation function in fully connected feedforward neural networks is the rectified linear unit (ReLU), which is defined as

$$f(x) = \max(0, x), \tag{3.6}$$

bringing all negative values to 0 (Fig. 3.6a) in contrast to the linear operator of identity $f(x) = x$.

The nonlinearity added by the action of bringing negative values to 0 has shown remarkable results in training ANNs. The "on-and-off" nature of ReLU mirrors biological neurons, which are either firing or inactive, and the total percentage of neurons that are firing at the same time is typically low at any given instant. In the same way, about half of the neurons using ReLU for their activation functions will be inactive. This sparse activation has been shown to improve training in many cases [17]. Furthermore, because ReLU is scale-invariant for positive inputs, it is not prone to the vanishing gradient problem.

Despite the broad success of ReLU in processing various types of data, it does suffer from several disadvantages which have prompted a number of variations to be developed. Firstly, and quite noticeably, the function is not differentiable at 0. This is typically addressed by arbitrarily setting the derivative to be either 0 or 1 at the point, or using instead a smooth approximation like the softplus function, whose derivative is the logistic function, though the former solution is faster to calculate. Secondly, the outputs of ReLU are not centered at 0, which means gradients can shift in a certain direction over time in some cases. While gradients do not shift towards 0, the network can run into a similar trouble called the "dying ReLU" problem, where neurons become inactive for almost all inputs, halting the training. Some alternatives to the ReLU include the leaky ReLU

$$f(x) = \begin{cases} x & \text{if } x > 0 \\ 0.01x & \text{else} \end{cases}, \tag{3.7}$$

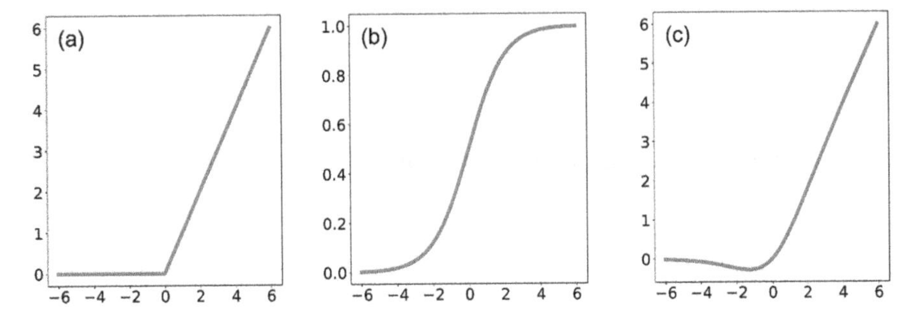

Fig. 3.6 Activation functions for (**a**) ReLU, (**b**) sigmoid, and (**c**) Swish

which shifts the negative values to $0.01x$ rather than 0, and a similar variant called the parametric ReLU

$$f(x)=\begin{cases} x & \text{if } x>0 \\ a\cdot x & \text{else} \end{cases}, \tag{3.8}$$

which shifts 0 to $a{\cdot}x$, with the coefficient a a parameter for each neuron that is learned through training. This leakage can effectively avoid the dying ReLU problem and has been shown to give better results in some cases.

Another common activation function is the SoftMax function:

$$f(x)_i=\frac{e^{x_i}}{\sum_{j=1}^{K} e^{x_j}} \text{ for } i=1,\ldots,K. \tag{3.9}$$

The SoftMax function takes in a K-dimensional input vector and normalizes it into a distribution where all values sum to 1. Because the outputs of the SoftMax function are in the range of (0, 1), it can be prone to the vanishing gradient problem and is thus not often used in the hidden layers. Instead, the SoftMax function is typically reserved for the final layer of ANNs used for multi-class classification problems, such that the output of the final layer can be interpreted as the probability of each of the possible classes. Two other activation functions, the hyperbolic tangent (tanh) and the sigmoid (or logistic) function (Fig. 3.6b), have been commonly used historically in training ANNs. However, both suffering from the vanishing gradient problem that leads to difficult optimization, their usage has shown a decreasing trend in recent years. One combination of these activation functions, called Swish and developed by Google, has been popularized recently. Figure 3.6c plots the Swish function, which is defined as the multiplication of the input x and a standard sigmoid function:

$$f(x)=x\cdot\frac{1}{1+e^{-x}}. \tag{3.10}$$

The Swish has been shown to outperform ReLU in certain particularly deep networks. Nonetheless, due to the obviously higher computational cost, it is not suitable for all tasks.

3.2.3 *Training of Deep Neural Networks*

The above subsections detailing the training process give the steps to pass one data sample through the ANN to update the weight parameters. For a dataset with tens of thousands of samples, this means tens of thousands of weight updates after feeding every single sample in the dataset, but given the high complexity of an ANN and all its possible weight configurations, this number of updates alone is not enough for most tasks. With a suitably conservative learning rate, each update is expected to push the network performance a miniscule amount towards improvement along the hyperspace of weight configurations. Since the initialization of weights is random, the starting point is likely to have extremely high error and be located a far distance from the actual global minimum. Thus, to continue reducing the error after all the data samples have been fed through, the same data need to pass through the network in more iterations, and a single iteration is called an epoch. Each epoch represents a full feedthrough of the training dataset. Many ANN training tasks could have hundreds to thousands of epochs, meaning each individual data point will be seen by the network hundreds to thousands of times. To illustrate why this is useful, consider the earliest couple of epochs in a simplified learning task. One of the earliest samples in the dataset might provide some unique critical information towards making a correct prediction, but because the weight initialization is so poor and noisy, the update will essentially focus on adjusting the weights to be less terrible; and if this sample is only seen once, the unique feature it carries may be wasted, whereas it may have provided a more useful update if it was seen at the end of training. At their core, weight updates revolve around strengthening connections that are helpful towards making an accurate prediction and weakening connections that are detrimental, but how useful this update is depends on what the current configuration of connections is. This explains the necessity and importance of feeding the same data repeatedly until the loss converges.

Feeding through each sample multiple times in repeated epochs is no doubt important, but there are also other factors to consider. While in theory, we would like each sample in the training dataset to provide some unique and valuable insight to assist the network in approaching optimal weights, in practice, the data are noisy, and the gradients from each individual sample have high variance. This means that some weight updates might contradict one other, and some may pull the loss function away from a minimum rather than towards it. If the dataset contains enough signal and is suitably designed, the loss function should still converge, but its path in the hyperspace of weight configurations may be chaotic and zigzagging, rather than smooth. Another factor to consider is the computational and time efficiency. Since each weight update involves calculating hundreds of thousands to millions of derivates, repeating the computation for all the samples and repeating the whole process for each epoch can quickly add up to an unwieldy total training time. One method to address these problems is called batch learning.

In batch learning, instead of running backpropagation and weight updates after each sample is fed through, an update is calculated only after all the samples are processed for a given epoch. This update will be an aggregate of the gradients from all the samples. The previous paradigm of feeding through samples individually and calculating updates for each one of them is called stochastic gradient descent, or online learning. There are a few notable advantages of batch learning over online learning. The first and most prominent one is the computational efficiency. Current popular software implementations of ANNs are extremely efficient in processing linear operations with large matrices in sublinear time. In other words, at each step along the backpropagation, calculating the gradients for a matrix encompassing all the samples takes less time than the time it takes to compute for a single sample multiplied by the size of the dataset. As a result, the most significant computational bottleneck in the training of ANNs is how many times the weights are updated, not how many gradients need to be calculated. In this regard, batch learning can provide a large reduction in the time consumed to conduct a single epoch of training. The second advantage is in reducing the previously mentioned instability from gradients of a single sample. Accumulating all gradients before an update of the learnable parameters significantly reduces variance and provides smoother and stabler updates.

The improved stability comes with some downsides, however. First, while each individual update is more likely to be accurate to the desired gradient towards the minimum, since only a single update is calculated per epoch, a larger number of epochs will be required to reach convergence. This can potentially offset the reduction in training time per epoch. Meanwhile, it also means that for individual training steps, the network learns slower but potentially more carefully. A second disadvantage is the purely negative treatment of noise and variance, which are sometimes beneficial. The hyperspace of the loss function is, in practice, never convex, and is riddled with many local minima and saddle points. The stable updates from batch learning can also make them more prone to get stuck at some local minima. Gradients of high variance provide the opportunity to knock the weight configuration out of where it gets stuck and onto a point where the gradients may lead to a lower local minimum. This is also why it is advised for training to start with a higher learning rate and decrease it over time. The initial weight configuration, being completely randomized, has little chance to be around the global minimum. Thus, more aggressive learning steps early on can help avoid higher error inflection points along the desired optimization path. A last disadvantage is one of practicality. When dealing with massive datasets, storing a matrix containing the entire dataset may have the hardware run out of memory.

To alleviate these concerns, a middle ground between batch and online learning becomes the most popular strategy in training ANNs, which is called mini-batch learning. Neither passing through the entire dataset nor a single sample at a time, the dataset is broken up into small chunks, called mini-batches, which are fed through in the fashion of one at a time, with the gradient being calculated as an aggregate of all samples only from the mini-batch. To make this clearer, for a total of N samples in the dataset and a mini-batch size of b, N/b weight updates will be calculated during each epoch (cf.: batch learning would have one update, and online learning would have N updates). Generally, the size of the mini-batches will be far smaller than that of the full training dataset and is specified by the user prior to training. The historical convention is based on factors of two, with the most common choices being 32, 64, and 128. The intuition behind is that CPU/GPU calculates the gradients using a memory architecture organized in powers of 2, potentially allowing for more efficient allocation of resources. Practically, however, the benefit of keeping batch sizes to strictly powers of two is almost negligible. As with the general trade-off between batch and online learning, larger mini-batch sizes offer a trade-off between shorter computation time and overly careful learning that may get stuck in undesired local minima. Finding the right batch size can be a process of trial and error depending on the task at hand, though only large changes to batch sizes, such as doubling or halving, will make noticeable differences to training. One can also take the same approach with dynamic learning rates to adjust the batch size over many epochs: smaller batch sizes are valuable in the beginning to encourage exploration of the hyperspace and avoid undesired local minima, whereas larger batch sizes are used towards the end as one is approaching the global minimum. Overall, mini-batch learning inherits some advantages from batch learning in computational efficiency and gradient stability as well as some from online learning in greater adaptability and learning speed.

3.2.4 Overfitting

One of the most significant pitfalls to watch out for when training ANNs, as with all ML models, is overfitting (Fig. 3.7). Any dataset is assumed to contain both useful information and noise, i.e., unwanted random fluctuations. The fundamental goal of any ML algorithm is to extract the signal that enables it to make accurate predictions when receiving independent data unseen in the training set. Again, if a model's task is to identify whether an image contains a cat, it will be trained on a set of labeled images with some of them containing cats. Then the end goal is, when given a brand-new picture for which the information about the presence/absence of cats is lacking, the model can have it correctly identified. Overfitting occurs when a model gets "overly powerful than necessary," such that it not only picks up the actual patterns in the data that are helpful to make correct predictions but also mistakenly treats the noise as useful information and relates it to the label. If this happens, the

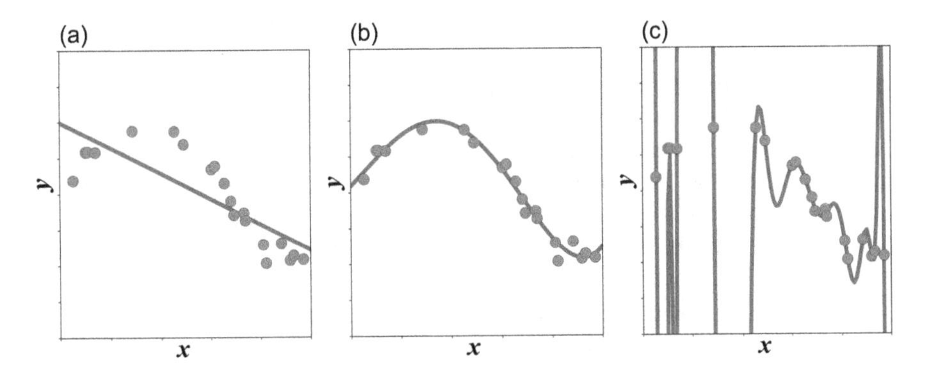

Fig. 3.7 (**a**) Data are underfit. Obvious discrepancies between the fit curve (in red) and training data points (in blue) indicate high bias error. (**b**) Data are properly fit. Despite some residual error, the model can be very generalizable to new data. (**c**) Data are overfit. The model treats noise as meaning patterns. Though it can match all the training data points well, the model is overly specific and has high variance, likely to generalize poorly on new data

model can classify perfectly on training data, but it would not generalize well, performing poorly on new data. Conversely, attempts at making the model more generalizable to new data could make it perform worse on the training data. This phenomenon is known as the bias-variance trade-off. ML models have both bias error and variance error. Bias is the error between the predicted and true values. High bias indicates underfitting, where the model is not learning enough of the useful information. Variance, on the other hand, is how sensitive the model is to small fluctuations in the data, or essentially how prone it is to overfitting by learning too much of the noise. The trade-off between these two sources of error is fundamental to the field of ML, and it can be shown that a completely unbiased model must have unbounded variance, and vice versa [18]. Therefore, a practical challenge of training ML models is managing and balancing bias and variance. Ideally, the errors on training data and on unseen new data should be very close and small, indicating good training and generalization performance of the ML model.

To assess how well a model will generalize on independent new data, there is no other way but to actually test it on new data. Therefore, for all ML tasks, it is recommended to have two datasets, a training set and a test set. We have mentioned training set several times in the above discussion, which is what is fed into the model through the training process. In common practices, the training set is larger than the test set, but the latter needs to have a substantial amount of samples as well. Most ML models will see the same data from this set over and over again during training as it is repeatedly fed through. Meanwhile, the test dataset is not used at all at any point during training. Instead, it is solely used in the post-training stage to evaluate how well the model performs on independent new data. In the case of ANNs, this means feeding the samples from the test set forwards to make predictions, comparing them to the corresponding labels to evaluate the network's accuracy, but not conducting the backpropagation to update weights. It is exceptionally important for users to watch out for data leakage, when data outside of the training set directly or indirectly affects training, as it can cause overfitting. The evaluation on the test set should only be conducted after the training is complete. If a user continually retrains a model while tuning its architecture or hyperparameters until a low test score is reached, it is essentially an indirect form of data leakage, because knowledge from the test data has been used since the first retraining. To this end, it is prudent to further split the training dataset into two: one set that is actually used for training and another validation set that serves as a proxy to estimate the model's performance on the test set. The validation set should be a random cross section of the data that is different every time the model is retrained, in order to prevent the over-tuning problem mentioned previously.

The extreme plasticity of ANNs, resulting from the massive learnable parameters to optimize, promises one of their greatest strengths as an algorithm but also makes them more susceptible to overfitting than most ML models. While the above descriptions on training and test sets introduce how one checks for overfitting, they tell little about how to avoid or alleviate the problem. There are a couple of methods for ANNs to address overfitting. First, according to statistical information theory, the upper bound of the expected difference between training and test errors grows with the capacity, or complexity, of the model but decreases with a larger number of training samples [18]. As such, having a larger training dataset can reduce the chance of overfitting. Intuitively, when more samples are used, the variance error tends to smooth out as opposed to what the actual patterns do. However, since obtaining labeled data is often a big hurdle, this simple solution is rarely practical, and the training time will increase accordingly. The second trick is changing the validation strategy. One popular choice is the k-fold cross-validation. In this method, the training data is randomly partitioned into k folds (or sections), and the same model is trained k times: In the first iteration, the first fold is used as validation, and the remaining $k - 1$ folds are used for training. Next, the second fold is used for validation while the remaining for a fresh round of training from scratch. The process continues for every fold left, and the average validation error across all folds is calculated at the end. Because the folds are random each time training is conducted, this method is quite robust for tuning models to prevent overfitting. In fact, it is rare for a model to perform well on cross-validation but poorly on the test set, if both sets are drawn from the same source and sufficiently large. Although running validation for k iterations increases the total training time, it ensures that a high validation score will not be obtained by chance, which is particularly important when the training dataset is small. Third, one can also choose to simplify the model architecture. If the task is not too complex, using a far deeper network than needed or having too many neurons per layer will make the model more liable to overfitting, since the excess connections will try to pick up additional patterns in the data. For a given problem, there are no hard and fast rules as to the right network complexity. Finding a suitable model architecture will typically require some trial and error and could be time-consuming.

Another class of approaches to avoiding overfitting is regularization, which refers to any modifications to a learning model which are intended to reduce its generalization error but not training error [3]. As mentioned above, the upper bound on the degree of overfitting decreases with a larger training set and increases with the model capacity. Regularization deals with the problem with a focus on the latter by reducing the complexity of a model. A versatile and computationally inexpensive strategy is the use of a technique called dropout. Dropouts are applied to specific layers of an ANN and have the network to randomly ignore some neurons during certain training steps. When a neuron is "dropped out," it outputs the value 0, essentially having no connections to the neurons in the previous layer. Dropout restricts the model's power in picking up patterns from the training data, implying weakened predictive capabilities and an increase in the training error, but at the same time makes it generalize better to new data. Another popular strategy, which often includes regularization explicitly in the naming, is the application of an extra penalty term to the loss function. Techniques using different penalties like L1 and L2 norms (of the weights) are thus named as L1 and L2 regularizations, respectively. The core idea is to force the network to only use as many connections as it absolutely needs and disregard others by setting their weights close to 0. Also worth mentioning is, for both strategies, applying regularization too aggressively can cause the network to abandon useful connections and lead to underfitting.

As a quick recap, all the methods discussed above on avoiding overfitting result in a worse performance on the training dataset, underlying the fundamental bias-variance trade-off. Managing this balance is one of the central concerns in training ANNs.

3.3 Variations of Neural Networks and Popular Models

A deep fully connected network has extraordinary power to model a wide variety of datasets. In fact, the universal approximation theorem states that a feedforward neural network with at least one hidden layer with any activation function can approximate any continuous function mapping from a finite-dimensional space to another, given that the network has enough neurons [3]. In practice, however, while standard feedforward networks can be applied to any problem, they may require an unwieldy amount of computational resources and/or achieve poor convergence. As such, several variations of network architecture have been developed over the years, which can differ in the number of connections between neurons, the way neurons are connected, and how values are transformed from layer to layer. These alternative architectures typically trade some of the raw predictive power and plasticity of fully connected NNs for a more focused and efficient network structure tailored to some specific types of problems and input data. In this subsection, we introduce two variations of NNs, namely, convolutional and recurrent NNs, and a few popular DL models, chosen based on their current appearance in photonics research.

3.3.1 Variations of Neural Networks

3.3.1.1 Convolutional Neural Networks

Convolutional neural networks (CNNs) are a type of NNs most applied to processing grid-like data such as images. Compared with other NN architectures, CNNs are distinguished by the use of convolutional layers, which execute a special linear operation called convolution. The main function of a CNN is to break down the input image into a series of features, from local ones extracted by the early layers to higher-level ones assembled in the later layers, and learn how to associate those features with the desired output [3, 4, 9]. This is reflected in the most common structure of CNNs, where convolutional layers for feature detection are arranged right after the input layer, followed by fully connected layers to learn the hidden relations within the features and produce the final output. For image data, each input image is represented by a tensor of pixel values in the shape of height × width × depth. The last dimension refers to the number of channels, which takes different values based on the color space, e.g., 1 for gray scale; 3 for red, green, and blue (RGB); and so forth. The following discussion assumes 1 channel for simplicity.

Convolutional layers work by sliding a filter of fixed size, or a convolution kernel, over the image and performing a convolution operation in each position (Fig. 3.8a). The filter itself is a matrix but typically in a much smaller size than the input image. As a result, a filter operates on a small window out of the total image, called the receptive field. The convolution produces a single value that is the sum of all the values in the receptive field weighted by the respective values in the filter. An ideal filter will be able to extract low-level features like edges, curves, and lines, and the numerical output of the convolution can be interpreted as a score of how well that particular part of the image in the receptive field matches the feature defined by the filter, i.e., how "edge-like" or "curve-like" that part of the image is. Once the convolution is computed, the window slides to a new portion of the image that becomes the next receptive field, and the image values therein are convolved with the filter values to produce another numerical output. This process is repeated so that the sliding filter covers the entire image, going left to right and top to bottom, where the user can predefine how far the receptive field moves each time, called the stride. The output of the current convolutional layer is another matrix, usually in a smaller size than the input image. There are two factors that affect the dimension of the output: stride and padding. Both are chosen prior to training as part of the hyperparameters. Stride, as just mentioned, controls how far the receptive field moves between each convolution; higher strides result in lower output dimensions. Padding adds extra pixel values around the perimeter of the input image. The main advantage of using padding is to adjust the output to a desired dimension (e.g., equal to the input). It also yields practical benefit if there are useful features at the boundary of the images. Overall speaking, for an $n \times m$ input image with an $a \times b$-sized filter, stride length s, and number of padding layers p, the final output dimensions are given by $[(n - a + 2p)/s + 1] \times [(m - b + 2p)/s + 1]$.

To give a simple example, the Modified National Institute of Standards and Technology (MNIST) database is a commonly used dataset for benchmarking NN implementations for image processing [19]. MNIST contains over 60,000 grayscale images of handwritten digits 0–9 in a fixed size of 28 × 28 pixels. Different images of the same digit may differ based on the handwriting style, so identification of a

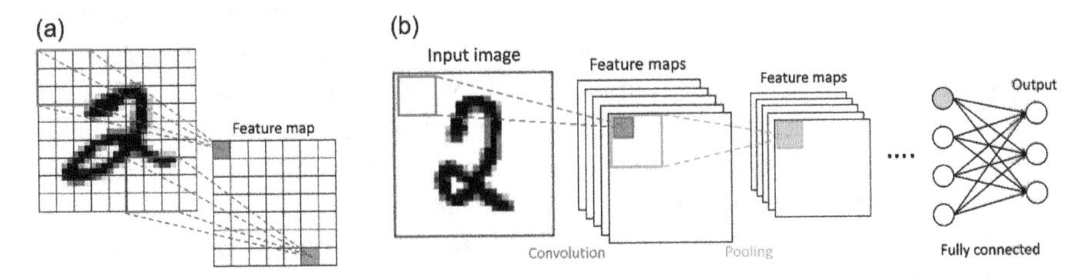

Fig. 3.8 Typical CNN architecture. (**a**) A convolutional layer slides a small filter (outlined 3 × 3 square) over an image, performing in each position a convolution between the filter values and the respective image values in the receptive field to generate a feature map. (**b**) A CNN uses alternating convolution and pooling layers to learn abstract features in the image, followed by fully connected layers to output the prediction. Nonlinearity applied to the feature maps after convolution is not shown in the diagram

digit requires the algorithm to recognize some characteristic patterns that help to differentiate between different digits. In the case where a 3 × 3 filter is used, the receptive field will start with the 3 × 3 area on the top left corner of the image and produce the first value, which is assigned to the top left grid in the next hidden layer. Then, with a stride length of 1, the receptive field will shift 1 pixel to the right for the next convolution computation. This process is repeated until pixels in the rightmost column are reached, and it will start over from the leftmost column but 1 pixel lower. As the filter runs over the entire image, the final output will be a 26 × 26 matrix, 2 pixels smaller in each dimension than the input, if conducted without padding. This output is essentially a new "image" that serves as a feature map, indicating how different parts of the input image match with the desired feature represented by the filter.

It is common to have multiple filters in a convolutional layer, because there are different useful features in every image, but each filter is associated with one particular feature. The additional filters operate in the exact same fashion, and the generated feature maps are stored to another channel. For example, with a convolutional layer that has ten filters, we will obtain ten feature maps from a 28 × 28 MNIST image and produce a 3D tensor of shape 26 × 26 × 10. Depending on the task, such procedures would be repeated with additional convolutional layers to extract higher-level features. Lastly, fully connected layers will be applied to the flattened feature maps to produce the final output. For example, a CNN trained on the MNIST dataset may learn to identify a semicircle, a diagonal line, and a horizontal flat line, and if those features are located on top of each other in the given order and proper orientations, the end result of recognition is highly likely to be a digit 2 (Fig. 3.8b).

Unlike the goal of learning in a fully connected layer which is to optimize the weights between each input and output neuron connection, the goal in a convolutional layer is to learn the filter matrix. Starting with randomized values in the filters, a CNN tries to learn what specific types of features help to reduce the loss. In theory, a fully connected layer can do everything a convolutional layer can. However, the main advantage of convolutional layers is the high computation efficiency [4]. CNNs prioritize the detection of local information, i.e., the relation of each pixel to its neighbors, to achieve sparse connectivity. For data where the relation between distant pixels is not typically meaningful, this saves a lot of computational resources for training. In the above example, where 28 × 28 pixels are mapped to a 26 × 26 matrix, a fully connected layer would have $(28)^2 \times (26)^2 = 529{,}984$ weights to optimize. In comparison, a single 3 × 3 filter, which is shared across all the positions as it slides over an image, only has nine trainable parameters. This massive reduction in the number of parameters allows convolutional layers to train more efficiently than similarly sized fully connected layers. Another route to improving computation efficiency is to follow each convolutional layer with a pooling layer. The latter operates by sliding another window over the feature map with predefined size and stride, but the output is either the maximum or average of the values in the receptive field. This down-sampling operation essentially summarizes the response of nearby pixels. For most image-based ML tasks where the location of high-level features being off by a few pixels will not significantly affect the classification, pooling effectively reduces the output dimension, further easing the computational requirements for the succeeding layers. Typically, a CNN stacks several segments, each containing one or a few convolutional layers and a pooling layer, and fully connected layers are added after the last pooling layer, as shown in Fig. 3.8b.

While all the above examples are detailed on 2D grayscale images, the same can be applied to input data of any dimension, whether those are 1D spectra, 3D color images with RGB channels, etc. For processing data containing spatial relationships between neighboring elements and between patterns, CNNs can offer a high predictive power and more efficient training.

3.3.1.2 Recurrent Neural Networks

Recurrent neural networks (RNNs) are a class of NNs specialized for processing sequential data [2]. Compared with feedforward networks discussed thus far, RNNs introduce a unique topology to the architecture, i.e., cyclical connections. In the simplest form, an RNN has a single, self-connected neuron in the hidden layer, as sketched in Fig. 3.9a. We will base the following brief discussion on this basic architecture. For advanced variations of RNNs (e.g., long short-term memory and gated recurrent units) as well as an emerging class of alternative NNs termed transformers, the interested reader is referred to other sources such as [4, 20, 21].

Enabled by the cyclical connections, RNNs can in principle take data sequences of arbitrary lengths and incorporate the inputs at earlier time steps into the computations, essentially remembering the past and using that information to help make better predictions. This is particularly useful for processing sequences where the context around an individual input element is just as important as the element itself and where an input could have multiple correct outputs depending on the preceding elements in the sequence. One area where RNNs are commonly used is natural language processing. In this application, data are sentences comprising a series of

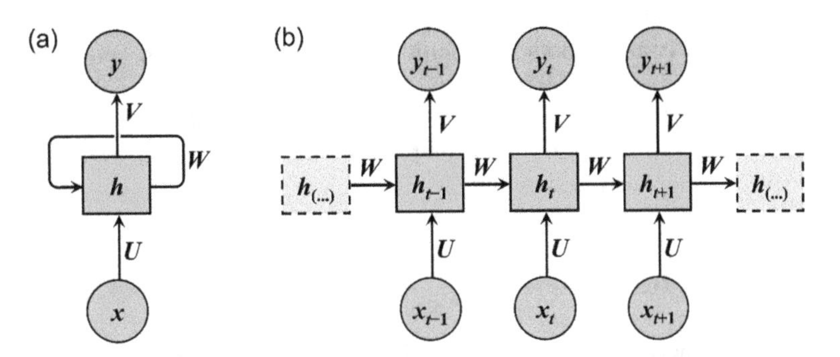

Fig. 3.9 Typical RNN architecture. (**a**) A recurrent neuron. A hidden state $\boldsymbol{h}$ is calculated based on the input $\boldsymbol{x}$ and the hidden state itself from a previous time step and is then used to calculate the output $\boldsymbol{y}$. $\boldsymbol{U}$, $\boldsymbol{V}$, and $\boldsymbol{W}$ denote the corresponding weight matrices, which are shared across all time steps. (**b**) Unfolded computational graph of the RNN cell in (**a**)

inputs, each an embedding of an individual word. For example, most smartphones incorporate a predictive text feature for messaging. When given a partially complete sentence, the app would suggest the potential next word so the users do not need to type it out. The most recently typed word is often not informative enough for intuiting the intended sentence. Adding even a single additional word can have important consequences. For example, "I can't" and "That can't" may be related to drastically different contextual information. Let us then take a quick look on how the architecture of an RNN is beneficial for processing sequences. Figure 3.9b illustrates the computational graph of an RNN cell, which is essentially Fig. 3.9a unfolded through time. A recurrent neuron maps from an input sequence $\boldsymbol{x}$ of arbitrary length $[x_1, x_2, \ldots, x_n]$ to an output sequence $\boldsymbol{y} = [y_1, y_2, \ldots, y_n]$ via a hidden state $\boldsymbol{h}$ that is continually updated at each time step $\boldsymbol{t}$:

$$h_t = f^{(h)}\left(W \cdot h_{t-1} + U \cdot x_t\right), \tag{3.11}$$

$$y_t = f^{(y)}\left(V \cdot h_t\right). \tag{3.12}$$

Here, $f^{(h)}$ and $f^{(y)}$ are the nonlinear activation functions. In Eq. (3.11), the first term in the parentheses contains information from preceding inputs, unless at the initial step where h_0 is from initialization. There are several algorithms for training RNNs. Among them, the backpropagation through time is commonly used and known to be efficient in computation time. Details of training methods can be found in some textbooks and monographs [22].

3.3.2 *Popular Deep Learning Models*

The above discussions have covered the major variations of NN structures. For many DL tasks, the models need to be tailored with more specifications and, therefore, built with more complicated architectures using the basic NN components (just like a CNN contains both convolutional and fully connected layers). In the following, we briefly discuss a few popular DL models. Although in ML textbooks they may be introduced in a different order to better reveal the ideas behind and connections between them [2, 3], we simply present them as a preparation for the later chapters of applications in photonics research.

3.3.2.1 Mixture Density Networks

Mixture density networks (MDNs) are a class of NNs that combine a deep neural network with a mixture model, a classic statistical model for representing a probability distribution with multiple distinct subpopulations. The aim of MDNs is to model multimodality with deep neural networks [23, 24]. In a conventional network, each neuron in the output layer corresponds to a discrete value for the desired prediction. While this framework seems to be straightforward, it requires an underlying assumption that there is only a single correct output for every input. If multiple identical or near identical inputs with differing outputs exist in the training dataset, the training can be significantly hampered, because different outputs will pull the weights in conflicting directions. Even if convergence is attained, due to the deterministic nature of the network, it may converge to the mean between those correct answers, not reflecting the true input-output mapping.

MDNs provide a solution for modeling multimodal data. Instead of using discrete values, it is also practical to describe the likelihood of a variable to take each of its possible values. Common probability distributions include the Gaussian distribution, Bernoulli distribution, Multinoulli distribution, etc. In order to model more complicated distributions and tackle multimodality, MDNs take one step further by employing mixture distributions, a straightforward yet powerful method to combine simple distributions. Specifically, MDNs use Gaussian distributions as the mixture components [3]. As a consequence, the neurons in the output layer of an MDN correspond to the parameters (π_k, μ_k, and σ_k) defining the mixture model [24]:

$$p(\boldsymbol{t} \mid \boldsymbol{x}) = \sum_{k=1}^{N} \pi_k(\boldsymbol{x}) \mathcal{N}\left(\boldsymbol{t} \mid \mu_k(\boldsymbol{x}), \sigma_k^2(\boldsymbol{x})\right). \tag{3.13}$$

Equation (3.13) describes a conditional density function, the probability of the modeled variable $\boldsymbol{t}$ given an input $\boldsymbol{x}$, where π_k, μ_k, and σ_k are the mixing coefficient, the mean, and the standard deviation of the k-th Gaussian component, respectively. In practice, the number of components in the mixture is predefined by the user as a hyperparameter. For an M-dimensional vector modeled by a mixture model with N components, the total number of output neurons is given by $(M + 2) \times N$, of which $M \times N$ is for the means. An example with two Gaussian components ($N = 2$) for a single variable ($M = 1$) is shown in Fig. 3.10. In most cases, unless the data contain many distinct subpopulations, a small number of components would suffice.

The training of MDNs needs some special considerations due to the change at the output layer. A different kind of loss function than one that directly compares the prediction and ground truth (such as MSE) is needed. As we are modeling a mixture Gaussian distribution, following the principle of maximum likelihood, negative log-likelihood can be used for the loss function. Working with the natural logarithm allows the loss (or, in this case, error) to be expressed in the common form of summation rather than product [23], alleviating data underflow issues. A trained MDN is still deterministic, always producing the same parameters for a given input. However, because the distribution is probabilistic, it introduces stochasticity. When a certain input has multiple possible correct answers, an ideally performing MDN would produce a probability distribution with peaks around each of those values, i.e., the correct answers are captured as modes in the distribution.

The application of MDNs can be quite diverse. One example is the linguistic features in speech synthesis, where the same text can be spoken in different ways

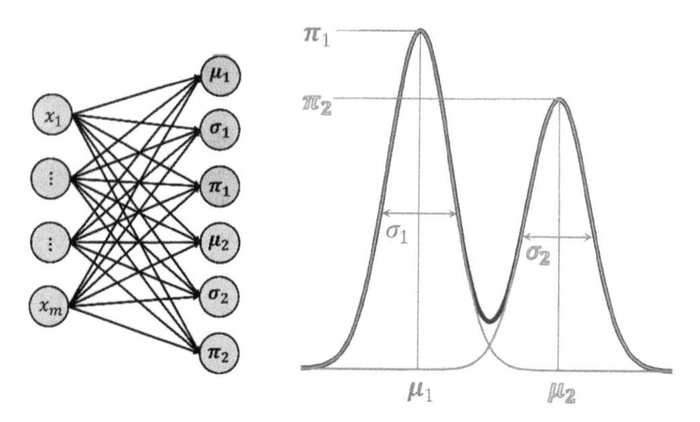

Fig. 3.10 In an MDN, the input vector $\boldsymbol{x}$ maps to a series of parameters of Gaussians, μ_k, σ_k, and π_k, which together form a mixture model. The shown example uses two Gaussian components ($k = 1, 2$) to model a single variable

[25]. Interestingly enough, MDNs have been used in the development of Siri, where it is also made recurrent [26]. In computer vision tasks, the recurrent MDN can be further combined with CNNs [27].

3.3.2.2 Generative Adversarial Networks

We now move on to generative models. While more rigorous definitions require statistical terms, we quote a concise one from [24]: "approaches that explicitly or implicitly model the distribution of inputs as well as outputs are known as generative models." Generative adversarial networks (GANs) are an important class of NNs in this category, and the framework was first proposed by Goodfellow et al. in 2014 [28] "for estimating generative models via an adversarial process." The goal of GANs is to generate synthetic data similar (in distribution) to the real data it is trained on. When it comes to image data, to which GANs are most often applied, this means that a model would create fake images indistinguishable from real ones. In terms of network architecture, a GAN consists of two separate neural network components that act to compete with each other. The first one is the generator, which maps vectors randomly sampled from a simple distribution (e.g., a Gaussian) as the input (also referred to as noise) to new candidate data as the output. The second component is the discriminator, which takes in data both from the training set (i.e., real data) and the samples created by the generator (i.e., fake data) and tries to tell whether it is real, a task of binary classification (Fig. 3.11). The goals of both networks during training are in direct opposition: The discriminator attempts to reduce the error when distinguishing between real and fake data, while the generator aims for the discriminator to have a higher error by producing data that look like they are from the training set. Accordingly, the training of GANs contains alternating steps

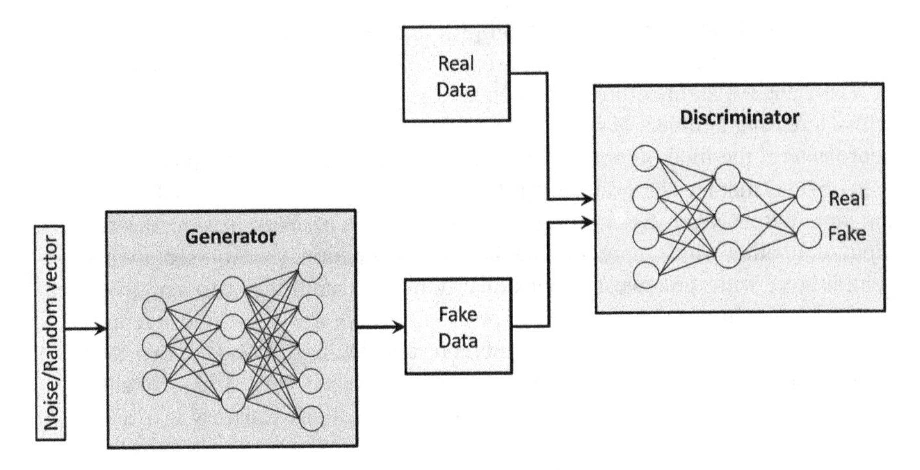

Fig. 3.11 Typical architecture of a GAN consisting of two NNs: a generator (left) and a discriminator (right). The generator takes in random input vectors to create fake data that look indistinguishable to the training data, and both fake and real data are fed into the discriminator, which aims to tell them apart. Fully connected layers are drawn for illustrative purposes. In practice, convolutional layers are often employed as well

of optimizing the discriminator and the generator [28], along with other tricks to address some technical difficulties [2].

The reader may have realized that for processing image data, the discriminator needs to be armed with convolutional layers to extract embeddings from the input. What deserves a few extra words is about the generator. As fake images are mapped from noise vectors, layers operating the opposite of convolution, namely, transposed convolutional layers [29], are employed to perform up-sampling. Through training, the generator learns the mapping from a simple distribution to the true data distribution. The generator of a trained GAN can be used to produce new data hopefully as realistic as the real ones.

GAN itself has many variations. The architecture introduced above is the vanilla version, where the input only has a random sample from a simple distribution. Additional information can be added to the generator input, e.g., the label of the image that we hope to generate—it can be concatenated with the random sample and together serve as the input. This is an example of the conditional GAN (cGAN), as we restrict what we want the generator to produce with some extra conditions (here the label of the image).

3.3.2.3 Autoencoders

Autoencoders are a type of NNs that learn to copy their inputs to their outputs [3]. This seemingly trivial task can be actually difficult in practice, as it requires finding a good way of efficient data representation. Traditional autoencoders were used primarily for dimensionality reduction [30] and feature detection [2], while some recent variations have turned generative and are used in computer vision tasks.

The standard architecture of autoencoders comprises a series of hidden layers with decreasing numbers of neurons, followed by layers with increasing numbers of neurons until the final output that has a dimension identical to the input layer. It is then obvious that the central layer has the fewest neurons—usually much fewer than the input/output layer. Because the goal of the entire network is to reconstruct the input or, in other words, approximate the identity function, the representation at this central layer with the lowest dimensionality must contain the most important features of the input data in a highly compressed form. In this sense, those representations are called codes (or codings, latent representations, etc.), and the two halves of the network are termed encoder and decoder, respectively (Fig. 3.12). Detailed discussion on more aspects of autoencoders in the traditional category is available in most textbooks.

Variational autoencoders (VAEs) are the best known type of autoencoder variations that serve as generative models [31]. The practical usage of VAEs, much like GANs, is to generate samples that resemble the ones from some desired dataset. When a standard autoencoder encodes data into a latent representation z, it places no restriction on the distribution of z, as its sole purpose is to reconstruct the input. However, this is not the case for VAEs. In the inference phase of a generative model, one can only sample noises from a known distribution. If such known distribution is too different from the one learned during training as the decoder input, no effective sampling of z can be obtained for the decoder to generate new data similar to $\boldsymbol{x}$. To match the learned distribution of z during training with such known distribution, the

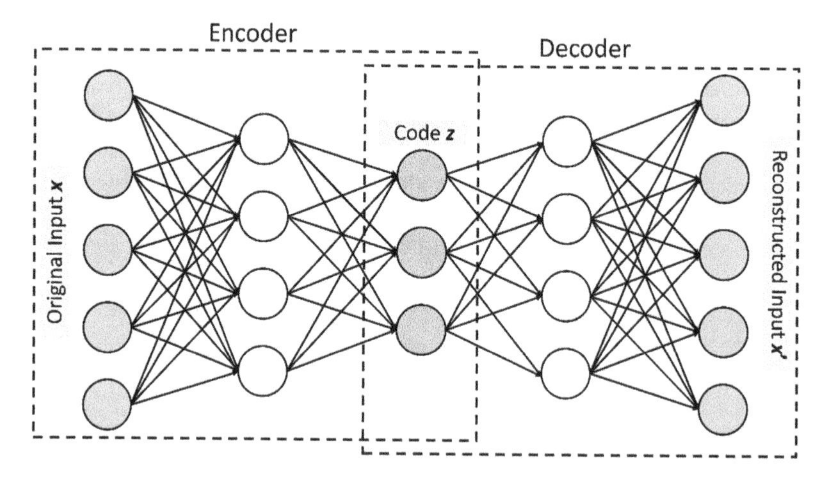

Fig. 3.12 Schematic of a standard autoencoder. The input and output layers are of the same dimension, with the smallest number of neurons in the central layer. The first half of the network, i.e., the encoder, compresses the input $\boldsymbol{x}$ into a code z in the latent space, and the second half, i.e., the decoder, is designed to reconstruct the input $\boldsymbol{x}$ with minimal loss by outputting $\boldsymbol{x}'$ from the code z

objective function of VAE uses an additional regularization term besides the standard reconstruction loss. Given two probability distributions P and Q over the same variable, their statistical distance can be measured by the Kullback-Leibler (KL) divergence, defined as

$$\mathrm{KL}\left(P \parallel Q\right) = \sum_{i \neq j} p_{ij} \log \frac{p_{ij}}{q_{ij}}. \tag{3.14}$$

The actual objective function for training VAEs is known as the evidence lower bound (ELBO), which differs from the KL divergence between the learned distribution of representation z and the known distribution by a negative sign and another term commonly used in other autoencoders [3]. Therefore, maximizing the ELBO is equivalent to simultaneously maximizing the reconstruction likelihood and minimizing the KL divergence.

3.4 Miscellaneous Machine Learning Techniques

3.4.1 Dimensionality Reduction

Data in ML are often high-dimensional, with each dimension comprising a feature useful for making predictions. ML algorithms tend to leverage their power most strongly with large datasets, both in terms of the number of samples and the number of features, which nonetheless can pose challenges. Besides the obvious concern over computational efficiency, a frequently discussed problem is the "curse of dimensionality" [2, 3, 32]. As the number of dimensions (features) in the data increases, the volume of the feature space grows drastically, and thus the data points become sparse. This can cause problems for many ML algorithms, particularly those for classification and clustering that are based on measuring the distances between data points, such as support vector machines, k-nearest neighbor classifiers, or k-means clustering. In theory, increasing the size of the training set could help to overcome the curse of dimensionality. However, this solution is not very practical, because when the number of dimensions increases, maintaining the same density of data requires exponentially larger numbers of samples. Alternatively, if an approach can reduce the number of dimensions (features) without sacrificing crucial information, it will offer significant advantages in understanding data and enabling ML algorithms. Reducing the data to two or three dimensions also helps for the purpose of visualization.

One of the most popular techniques for dimensionality reduction is principal component analysis (PCA). PCA transforms a dataset into an orthogonal coordinate system, where the space is spanned by a set of vectors called principal components (PCs) [33]. In essence, PCs are the linear combinations of vectors from the original coordinate system, and they are constructed such that along their directions, the maximum amounts of variance in the data are preserved. Specifically, the first PC contains the largest amount of variance in the dataset, the second contains the next most while being orthogonal (i.e., linearly uncorrelated) to the existing component, and so on (Fig. 3.13a). In this way, the most useful information in a high-dimensional dataset can be concentrated into a reduced number of dimensions: one only needs to project it onto a subspace defined by the first few PCs, whereas dropping the remaining components will likely not cause much loss of important information.

To construct the principal component coordinate space, first the covariance matrix of the data points is calculated. For a dataset X comprised of N feature vectors $[X_1, X_2, \ldots X_N]$, the covariance matrix K_{XX} is

$$K_{XX} = \begin{pmatrix} \operatorname{cov}(X_1, X_1) & \cdots & \operatorname{cov}(X_1, X_N) \\ \vdots & \ddots & \vdots \\ \operatorname{cov}(X_N, X_1) & \cdots & \operatorname{cov}(X_N, X_N) \end{pmatrix}, \tag{3.15}$$

where the covariance of a variable with itself, along the diagonal, is the variance of that feature vector. The covariance between different variables describes how one variable would change in response to a change in the other, such as whether they are directly or inversely correlated. PCs are then computed by doing the eigen-decomposition of the covariance matrix. The eigenvectors of K_{XX} are the PCs, and the corresponding eigenvalues represent the amount of data variance contained in each

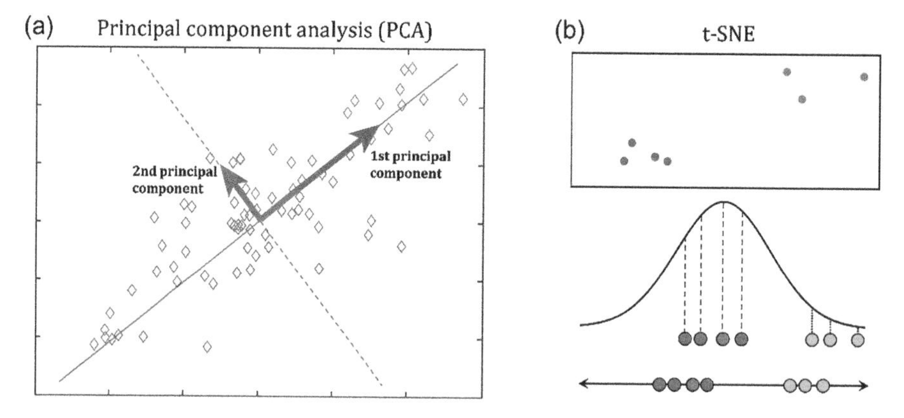

Fig. 3.13 (**a**) PCA on a 2D dataset. The first PC is along the direction which contains the largest amount of variance when the data are projected onto it, and the second PC is orthogonal to the first. (**b**) Diagram of simplified t-SNE from 2D (top) to 1D (bottom). A probability distribution models the points based on their proximity to each other, and the lower-dimensional embedding seeks to preserve such relative distance and place similar points close and dissimilar points far

PC. Therefore, in deriving the final matrix of PCA, the eigenvectors can be well ordered through ranking their respective eigenvalues, allowing the user to optimize their choice that to what degree the dimensionality can be reduced.

Another popular method for dimensionality reduction is *t*-SNE, which, differing from PCA, is a nonlinear algorithm [34]. The goal of *t*-SNE is to embed a high-dimensional dataset in a 2D or 3D space while having the relative distance between data points preserved, i.e., similar samples tend to be located near one another, and dissimilar samples are more likely to be far apart. Given a set of *N* data points $[x_1, x_2, \ldots, x_N]$ in a high-dimensional space, *t*-SNE employs the conditional probability $p_{j|i}$ to represent the similarities of point x_j to point x_i, which describes the chance that x_j would be chosen as the neighbor of x_i if such selections were made based on a Gaussian distribution centered at x_i (Fig. 3.13b). The mathematical expression of $p_{j|i}$ is

$$p_{j|i} = \frac{\exp\left(-\left\|x_i - x_j\right\|^2 / 2\sigma_i^2\right)}{\sum_{k \neq i} \exp\left(-\left\|x_i - x_k\right\|^2 / 2\sigma_i^2\right)}, \tag{3.16}$$

where σ_i is the variance of the Gaussian. The conditional probability is used in stochastic neighbor embedding (SNE) but nonetheless leads to some limitations [34]. To overcome those problems, in its successor, t-SNE, pairwise similarities p_{ij} based on joint probabilities are used, defined as

$$p_{ij} = \frac{p_{j|i} + p_{i|j}}{2N}. \tag{3.17}$$

Likewise, in the low-dimensional space where the data points are mapped to $[y_1, y_2, \ldots, y_N]$, the pairwise similarities q_{ij} are given by

$$q_{ij} = \frac{\left(1 + \| y_i - y_j \|^2\right)^{-1}}{\sum_{k \neq l} \left(1 + \| y_k - y_l \|^2\right)^{-1}}. \tag{3.18}$$

Equation (3.18) is defined based on a Student's *t*-distribution rather than a Gaussian. In brief, this choice is made to handle the distant data points better. For more detailed justification, we direct the interested readers to the original work [34]. The distance between the probability distributions p_{ij} and q_{ij} is measured by the KL divergence in Eq. (3.14). By minimizing the KL divergence, the representation of the points in the low-dimensional space matches the similarity relations of the data points in the high-dimensional space.

Although not obvious from the procedure described above, *t*-SNE has a few hyperparameters that can be tuned to adjust the spread of the compressed data, and bad choices of parameters can result in low-quality clustering. In contrast, PCA will produce a deterministic decomposition. Despite this, the nonlinear nature of *t*-SNE holds the promise for dimensionality reduction of more complicated datasets that may be linearly inseparable. Moreover, because t-SNE is specifically designed for compression down to two or three dimensions, it is often a better choice for visualization of data where a PCA reduction may have a non-trivial amount of information contained past the third principal component.

3.4.2 Transfer Learning

As powerful as NNs can be for certain problems, they are not by their nature very adaptable. NNs, like all standard ML algorithms, are, at their core, an optimization algorithm, and one that is data-driven rather than purely algorithmic. NNs are built with the assumption that the data they are trained on are representative. Solving any new problem that does not match the training data distribution requires both new data and a new model, which can be expensive on many occasions and inefficient if training the model from scratch.

In real-life situations, humans can use intuition and reasoning to adapt the knowledge needed for solving one problem to solving another. Transfer learning covers a series of methods to translate a similar idea to the realm of ML [3]. As its definition, categorization, and applications have been covered by several comprehensive surveys [35, 36], here we only explain the concept briefly. The weight parameters stored through training of a NN constitute a form of learned knowledge, and the goal of transfer learning is to convey that knowledge to related problems. For the simplest case possible, imagine adding new input variables to a task which already has a trained NN. If the new variables indeed contain useful information for making predictions, their inclusion should be solely beneficial. However, the previous NN only accepts input vectors in a fixed length, and with the new input variables, it no longer fits. This necessitates a different neural network with more neurons in the input layer and many more weights for all the new connections this will create. Training the new model from scratch with the reconfigured dataset is just straightforward, but one may feel something is not quite right, because all we want to address is adding new input variables to a NN whose training cost has been paid. Transfer learning can be helpful in this situation. A NN breaks a problem down into various levels of abstraction through its hidden layers. Some of them, as the manifestations of certain knowledge, could be reused in another NN to incorporate new information. Compared with training a new model from scratch, which usually has a high initial loss, a transfer learning approach is more likely to start from a better initialization.

A more common use case is transfer learning between different but similar types of tasks. For example, the knowledge gained in a model for identifying cats in images can be used for another model for identifying tigers. A CNN trained on the former problem typically devotes its early convolutional layers to learning the important spatial features about cats, such as which parts of the image correspond to ears, eyes, and tails. Later layers may focus on learning the correct orientations of those features relative to each other, such as how far apart the eyes should be, where the end of the tail can possibly be located, and so on. When it comes to training a new model for identifying tigers, likely a considerable number of features that have helped in the first task are still useful, owing to the visual similarities between the two animals. Meanwhile, the orientation information may need to be reevaluated due to tigers being much larger than cats and many other subtle differences. Consequently, the knowledge learned in the lower layers can be transferred with minimal adjustment, and that in later layers will still be far closer to the ground truth than a random initialization. Taking advantage of the knowledge from a related task (of cat identification), it is reasonable to expect the model for identifying tigers to be trained much faster, and possibly with a better optimum.

3.4.3 Reinforcement Learning

Reinforcement learning is a unique type of ML separate from supervised and unsupervised learning. While the latter two have been described under the umbrella of AI, the actual working mechanisms of both techniques are far from what the general public would think of when they hear the term: an artificial entity able to "think" like a human and make reasoned, intelligent choices. RL, on the other hand, fits a little better into that notion, as it involves training an agent to make decisions based on some predefined reward measure [2]. Some real-world applications are robot control and self-driving cars.

A typical RL problem contains a few key components and concepts [37]. First is an agent. This is the part of the algorithm that acts as a discrete "intelligent entity". The agent has access to a set of predefined actions [a_1, a_2, a_3, ... a_n] in the action space A. Second, an environment is defined, which is where the agent is put in and consists of a set of states [s_1, s_2, s_3, ... s_n] in the state space S. In the maze navigation problem, for example, the states are all the squares in the grid that the agent can occupy, the current state is where the agent stands, and the actions would be moving in any direction allowed. The next component is the policy. The policy is an algorithm, for example, a NN, the agent uses to map its current state to probabilities of choosing each possible action. There also need to be a system of feedback, called reward, the agent receives from the environment after each action, and subsequently, a value, which is a function of states that the agent uses to estimate its future reward when taking a certain action in a given state. Rewards are numerical values. Positive/negative rewards are often assigned to desired/undesired actions for completing the task. In RL, the objective of the agent is to learn a mapping from the intermediate states and values to an optimal sequence of actions so that its expected cumulative reward can be maximized. It is worth noting that an agent may accept negative rewards in the short term if that ultimately helps to maximize the expected reward over the long term.

The agent-environment interaction in RL can be summarized as follows (Fig. 3.14). First, the agent observes its current state s_t. Then, using the estimate of

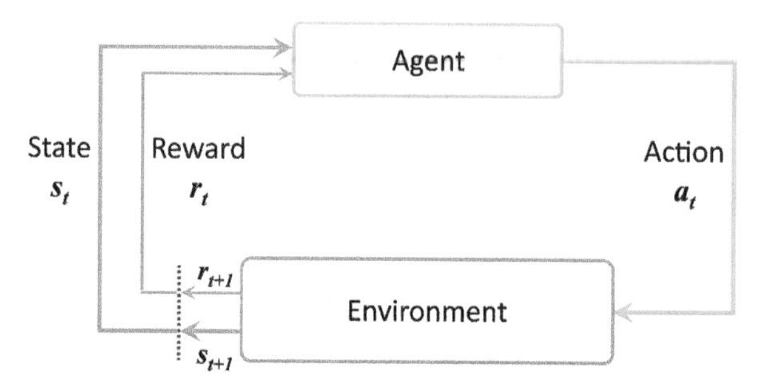

Fig. 3.14 Flowchart of reinforcement learning process. The agent observes its current state s_t and, using its estimates of values and its policy, chooses an action a_t. The environment proceeds to the next state s_{t+1} and gives the agent a reward r_{t+1}, and the agent uses them to update its value estimates. In a single episode, this process is repeated until the terminal state is reached

the value and its policy, the agent chooses an action a_t to take. In the next time step, the agent moves to the state s_{t+1} and gets an immediate reward r_{t+1} from the environment, based on which the agent updates and improves its internal model of the policy. This process is repeated until some terminal state is reached. The whole sequence constitutes what is called an "episode", essentially a single run through the given scenario. The training revolves around a series of episodes, such as different runs through the same maze, during which the agent learns more and more towards the optimal solutions to maximize the cumulative reward over time. It is easy to see that, during early episodes, the agent has no useful information and will act randomly until it hits the goal by pure chance. However, it will reinforce the states and actions that led to high rewards and gradually improve in the later episodes.

For cases with a finite number of states and actions, it is common for the agent to store an internal table of all the possible states $V(s)$ and/or state-action pairs $Q(s,a)$, which is filled with the agent's estimates of the values and will be continually updated through the training process as the agent receives more feedback. The policy may be for the agent to always choose the action that has the highest estimated value, which is called a greedy policy; or it may occasionally choose a random action with some probability ε to test if that can be more optimal, called ε-greedy. Generally speaking, a good policy has to make trade-offs between "exploitation," meaning that using the information it has learned from the past to make decisions, and "exploration," which favors choosing actions that currently seem to be suboptimal, in order to gain more information and improve the accuracy of value estimation. In the early stages of training, the agent has little information about the system, so exploration is more important.

There are many algorithms for reinforcement learning [38], such as Q-learning, which goes through episodes and iteratively update value tables based on the difference between the expected and actual outcomes, and temporal difference learning which uses bootstrapping. For a given scenario, most common RL algorithms can be proven to reach optimality eventually, although the speed and route could vary widely. Some algorithms also have versions incorporating other types of ML models, primarily deep neural networks, which are used to estimate the state and action values in lieu of tables.

References

1. Alpaydin, E.: Introduction to Machine Learning, 3rd edn. MIT Press, Cambridge (2014)
2. Géron, A.: Hands-on Machine Learning with Scikit-learn, Keras, and TensorFlow: Concepts, Tools, and Techniques to Build Intelligent Systems. O'Reilly Media, Inc., Sebastapol (2019)
3. Goodfellow, I., Bengio, Y., Courville, A.: Deep Learning. MIT Press, Cambridge (2016)
4. Kelleher, J.D.: Deep Learning. MIT Press, Cambridge (2019)
5. McCulloch, W.S., Pitts, W.: A logical calculus of the ideas immanent in nervous activity. Bull. Math. Biophys. **5**(4), 115–133 (1943)
6. Hebb, D.O.: The Organization of Behavior: A Neuropsychological Theory. John Wiley & Sons, New York (1949)
7. Rosenblatt, F.: The perceptron: a probabilistic model for information storage and organization in the brain. Psychol. Rev. **65**(6), 386-408 (1958)
8. Ivakhnenko, A.G.: Polynomial theory of complex systems. IEEE Trans. Syst. Man. Cybern. **SMC-1**(4), 364–378 (1971)
9. Fukushima, K.: Neocognitron: a self-organizing neural network model for a mechanism of pattern recognition unaffected by shift in position. Biol. Cybern. **36**(4), 193–202 (1980)
10. Rumelhart, D.E., Hinton, G.E., Williams, R.J.: Learning representations by back-propagating errors. Nature. **323**(6088), 533–536 (1986)
11. Krizhevsky, A., Sutskever, I., Hinton, G.E.: ImageNet classification with deep convolutional neural networks. In: Proceedings of the 25th International Conference on Neural Information Processing Systems – Volume 1, pp. 1097–1105. Curran Associates Inc, Red Hook, NY (2012)
12. Silver, D., et al.: Mastering the game of Go with deep neural networks and tree search. Nature. **529**(7587), 484–489 (2016)
13. Pedregosa, F., et al.: Scikit-learn: machine learning in Python. J. Mach. Learn. Res. **12**, 2825–2830 (2011)
14. Abadi, M., et al.: Tensorflow: Large-scale machine learning on heterogeneous distributed systems. arXiv preprint arXiv:1603.04467 (2016)
15. Gulli, A., Pal, S.: Deep Learning with Keras: Implementing Deep Learning Models and Neural Networks with the Power of Python. Packt Publishing Ltd, Birmingham (2017)
16. Paszke, A., et al.: PyTorch: an imperative style, high-performance deep learning library. In: Proceedings of the 33rd International Conference on Neural Information Processing Systems – Article No. 721, pp. 8026–8037. Curran Associates Inc, Red Hook, NY (2019)
17. Glorot, X., Bordes, A., Bengio, Y.: Deep sparse rectifier neural networks. In: Proceedings of the fourteenth international conference on artificial intelligence and statistics. Proc. Mach. Learn. Res. **15**, 315–323 (2011)
18. Hastie, T., et al.: The Elements of Statistical Learning: Data Mining, Inference, and Prediction, 2nd edn. Springer, New York (2009)
19. The MNIST database of handwritten digits
20. Graves, A.: Supervised Sequence Labelling with Recurrent Neural Networks. Springer (2012)
21. Vaswani, A., et al.: Attention is all you need. In: Proceedings of the 31st International Conference on Neural Information Processing Systems – pp. 6000–6010. Curran Associates Inc, Red Hook, NY (2017)
22. Chauvin, Y., Rumelhart, D.E. (Eds.): Backpropagation: Theory, Architectures, and Applications. Psychology Press, New York (1995)

23. Bishop, C.M.: Mixture Density Networks. Aston University, Birmingham (1994)
24. Bishop, C.M.: Pattern Recognition and Machine Learning. Springer, New York, NY (2006)
25. Zen, H., Senior, A.: Deep mixture density networks for acoustic modeling in statistical parametric speech synthesis. In: 2014 IEEE International Conference on Acoustics, Speech and Signal Processing (ICASSP) (2014)
26. Deep Learning for Siri's Voice: On-device Deep Mixture Density Networks for Hybrid Unit Selection Synthesis. Available from: https://machinelearning.apple.com/research/siri-voices (2017)
27. Bazzani, L., Larochelle, H., Torresani, L.: Recurrent mixture density network for spatiotemporal visual attention. arXiv preprint arXiv:1603.08199 (2016)
28. Goodfellow, I., et al.: Generative adversarial nets. In: Proceedings of the 27th International Conference on Neural Information Processing Systems – Volume 2, pp. 2672–2680. MIT Press, Cambridge, MA (2014)
29. Dumoulin, V., Visin, F.: A guide to convolution arithmetic for deep learning. arXiv preprint arXiv:1603.07285 (2016)
30. Hinton, G.E., Salakhutdinov, R.R.: Reducing the dimensionality of data with neural networks. Science. **313**(5786), 504–507 (2006)
31. Kingma, D.P., Welling, M.: Auto-encoding variational Bayes. arXiv preprint arXiv:1312.6114 (2013)
32. Bellman, R.: Dynamic programming. Princeton University Press (1957)
33. Jolliffe, I.T., Cadima, J.: Principal component analysis: a review and recent developments. Philos Trans R Soc A. **374**(2065), 20150202 (2016)
34. Van der Maaten, L., Hinton, G.: Visualizing data using t-SNE. J Mach Learn Res. **9**(86), 2579–2605 (2008)
35. Pan, S.J., Yang, Q.: A survey on transfer learning. IEEE Trans Knowl Data Eng. **22**(10), 1345–1359 (2010)
36. Torrey, L., Shavlik, J.: Transfer learning. In: Handbook of research on machine learning applications and trends: algorithms, methods, and techniques, pp. 242–264. IGI Global (2010)
37. Sutton, R.S., Barto, A.G.: Reinforcement learning: an introduction. MIT Press (2018)
38. Szepesvári, C.: Algorithms for reinforcement learning. Springer, Cham (2010)

Chapter 4
Deep-Learning-Assisted Inverse Design in Nanophotonics

Abstract Of the several possible ways in which nanophotonics and machine learning could blend, inverse design of nanophotonic structures enabled or assisted by ML algorithms is probably the most straightforward one. Historically, the design of optical and photonic devices is intuition-based, starting with deriving prototypes from some known physical principles and followed by manually tuning a handful of variables, which define the prototypical structures, to optimize the performance. We have seen examples of this type in the previous discussions on, for instance, optical antennas and SRRs. Nonetheless, as the demands for better device performance, such as higher efficiency, smaller footprint, and novel functionalities, keep increasing rapidly, the involved variables and in turn complexity of the design tasks explode, posing a substantial challenge for intuition-based design strategies to tackle. Computer programs have been introduced to the optimization of circuits and electromagnetic devices since the 1960s (Temes, Calahan, Proc IEEE 55(11):1832–1863, 1967; Cheng et al, Int J RF Microw Comput Aided Eng 20(5):475–491, 2010; Weile, Michielssen, IEEE Trans Antennas Propag 45(3):343–353, 1997; Vai, Prasad, IEEE Microw Guided Wave Lett 3(10):353–354, 1993), if not earlier. In photonics, the application of evolutionary and gradient-based algorithms to optimization problems can be tracked back to at least the late 1990s (Molesky et al, Nat Photonics 12(11):659–670, 2018; Spuhler et al, J Lightwave Technol 16(9), 1680–1685, 1998; Dobson, Cox, SIAM J Appl Math 59(6):2108–2120, 1999). The usage of optimization algorithms in inverse design has dramatically enhanced researchers' capability to explore the enormous design possibilities beyond traditional devices built upon elements in regular shapes and simple arrangements. ML, particularly DL, provides an alternative route of design optimization to novel nanophotonic devices in demand (Yao et al, Nanophotonics 8(3):339–366 (2019); Wiecha, et al, Photonics Res 9(5):B182–B200, 2021; Jiang et al, Nat Rev Mater 6(8):679–700, 2021; Ma et al, Nat Photonics 15(2):77–90, 2021). Compared to conventional optimization techniques, which by following predefined search rules approach the final design from an initial guess stepwise through

The original version of this chapter was revised. The correction to this chapter is available at https://doi.org/10.1007/978-3-031-20473-9_7

K. Yao, Y. Zheng, *Nanophotonics and Machine Learning*, Springer Series in Optical Sciences 241, https://doi.org/10.1007/978-3-031-20473-9_4

iterative cycles of simulation and fitness evaluation, DL-based design methods utilize the data from many simulations in advance and iteratively feed them into the DL model to improve its performance. This optimization process is essentially the training we talk about in the last chapter.

Concerning the necessity of using DL for photonic inverse design and the performance of this method in comparison to that of advanced optimization techniques, there are no simple answers in a general sense. Nonetheless, some critical comparisons have been made in the literature, e.g., in [1, 2]. On the one hand, it is important to keep in mind that when the same simulation tool is deployed for data generation and fitness evaluation, data-driven design methods cannot outperform iterative optimizations without a constraint of time. Additional errors are inherent to ANNs. For a not very large number of design tasks, it is typically much more efficient to approach better designs by running additional iterations of optimization than spending the same computation power on supplementing the training data [1]. On the other hand, DL-based design methods do have the advantage that once fully trained, ANNs can produce designs almost instantaneously, usually in the order of a few milliseconds, which is negligible compared with the time needed for completing even a single simulation. Therefore, for applications which involve heavily repetitive design tasks based on a common dataset, DL-based design methods can still serve as useful tools alternative to optimization techniques.

In this chapter, we discuss the application of DL to inverse design of nanophotonic structures and devices. The main purpose is to bring the reader a brief idea about one possibility that how the basic concepts introduced in the preceding chapters from two distinct disciplines can be connected. For the latest progress in this research direction as well as the utility of sophisticated DL models, we refer interested readers to some excellent review articles [1–6]. We would also like to remind the reader that, although not among the foci of this chapter or this book, advanced nanophotonic optimization techniques such as topology optimization (TO) have been critically discussed, reviewed, or systematically introduced in several classic works [7–10].

4.1 Inverse Design and Non-uniqueness Problem

We start by formulating nanophotonic inverse design. A nanophotonic structure or device can be associated with two types of characteristics by nature [2]. The first type contains variables that define the structure. The geometry and material properties of any components and the configuration of the incident light are the most common variables, known as design parameters. The second type is formed by variables which describe the optical responses of the structure. Examples of this type, which could vary across different applications, include the transmission (or reflection, absorption, scattering, extinction, etc.) behaviors or propagation properties (i.e., band structures [11]) over a certain wavelength range, the local field distributions or quantities derived from them (e.g., optical chirality [12] and nonlinear signals [13]), and far-field characteristics like radiation/scattering patterns, to name a few

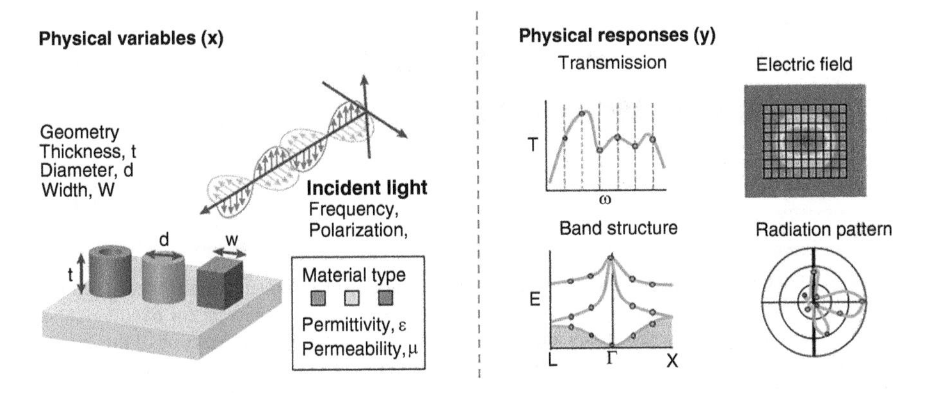

Fig. 4.1 Design parameters (left) and optical responses (right). Left: A nanophotonic structure can be described by parameters ($\boldsymbol{x}$) in several categories, including geometry such as the dimensions of individual building blocks and layout of their assembly, the material properties of each element and the environment, and configurations of the incident light such as frequency, polarization, and propagation direction. Right: The optical response ($\boldsymbol{y}$) of a nanophotonic device, depending on its functionality, can be expressed by a spectrum, a near- or far-field distribution, a band structure, etc. Computing the responses $\boldsymbol{y}$ of a device described by given parameters $\boldsymbol{x}$ is known as the forward problem, whereas retrieval of $\boldsymbol{x}$ from $\boldsymbol{y}$ is the inverse problem. (Reprinted from [2] with permission from Springer Nature)

(Fig. 4.1). For the convenience of discussion, hereafter we denote design parameters by the variable $\boldsymbol{x}$ and optical responses by $\boldsymbol{y}$, and they each span a space termed design space and response space, respectively.

It might be trivial but still necessary to point out that there exists a mapping between design parameters $\boldsymbol{x}$ and optical responses $\boldsymbol{y}$. For any given structures and excitation conditions, one can always determine the responses of interest by simulations and/or measurements, a process of solving the *forward problem*. Accordingly, inverse design is to solve the *inverse problem*, which is defined in the reverse direction and seeks to find structures for known or desired responses. The history of applying DL to solving electromagnetic problems is just as long as that with optimization techniques. Early attempts were made in the 1990s for microwave devices [14–21], followed by some for photonic crystals [22–24] (Fig. 4.2). The recent interests from the nanophotonic community are largely enabled by the tremendous development of DL and computation power during the past decade [2]. Essentially, DL is applicable to solving both forward and inverse problems. As the reasoning of its usage in inverse design has been explained in the introduction, now we only take a quick look at the forward problem. Traditionally, unless in exceptional cases where analytical solutions can be derived, the optical responses of a structure are modeled by using various simulation and numerical methods, such as finite-difference time-domain (FDTD), finite element method (FEM), boundary element method (BEM), finite integration technique (FIT), coupled dipole approximation (CDA), rigorous coupled wave analysis (RCWA), etc. These methods solve Maxwell's equations by discretizing the structures and their responses spatially and temporarily, resulting in a huge number of nodes and associated degrees of freedom that are computationally costly. DL is useful in this case, because once a model

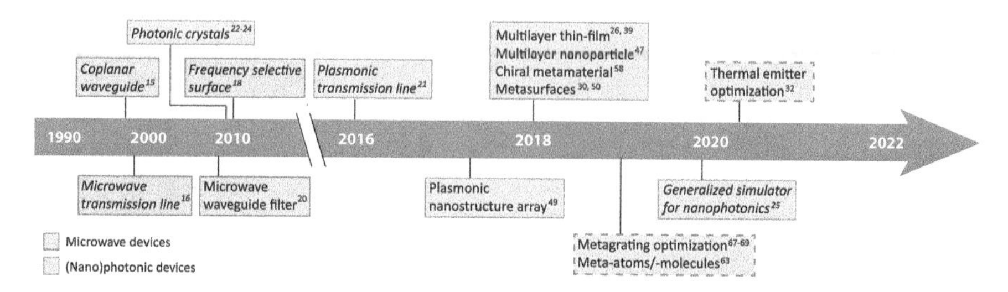

Fig. 4.2 Timeline of forward modeling and inverse design of electromagnetic devices based on ANNs, starting in the 1990s for microwave systems and running to the present for nanophotonic devices. Focus is given to recent years and on nanophotonic devices within the scope of this book. Items in regular/italicized text denote the use of ANNs for solving inverse/forward problems. Boxes with dashed outlines indicate hybrid design frameworks combining DL and optimization methods

learns the mapping from $\boldsymbol{x}$ to $\boldsymbol{y}$ for a group of structures, it can predict the responses much faster and serve as a surrogate simulator [25]. Note again that in terms of accuracy, DL models cannot outperform the original simulation method used to generate the training data [1].

Training a DL model as a surrogate simulator is typically much easier than training one for inverse design. The key factor determining this difference is the so-called non-uniqueness problem in the mapping from optical responses to design parameters. When solving a forward problem, the results are determinant. Every structure uniquely corresponds to one certain response. However, it is not rare that the same response comes from multiple different structures. To see this, let us consider a simple example of light scattering by nanoparticles. When the scattering spectra from two particles, which could differ in size, shape, material, or other properties, are overlaid, if there is any intersection between the two spectra, at that wavelength the scattering intensity has at least two possible sources. This one-to-many mapping or degeneracy of solutions causes confusion of standard ANNs during training. The consequence is the models will either have difficulty in converging or learn incorrect mapping between $\boldsymbol{x}$ and $\boldsymbol{y}$. One may argue that the two spectra can still be considerably different if the comparison is made over a wavelength range, not just at a single wavelength. Indeed, the above example is extreme and oversimplified. Optical responses in the form of high-dimensional data are less likely to totally overlap for different structures. Nonetheless, in most problems in nanophotonics, the design space and response space are both vast enough, where cases of very close optical responses coming from distinctly different structures often exist.

A variety of methods have been proposed to tackle the non-uniqueness problem in inverse design [1, 2]. Figure 4.3 presents a nice illustration summarized by Wiecha and coworkers [1]. For the ease of understanding, we reform the above example slightly in Fig. 4.3a, where the optical response is the extinction over a narrow range of wavelengths, and the design parameter is specified to the length of a gold nanorod. On the right panel, it is obvious that the mapping from extinction to the rod length (blue curve) is multivalued. As mentioned earlier, standard ANNs

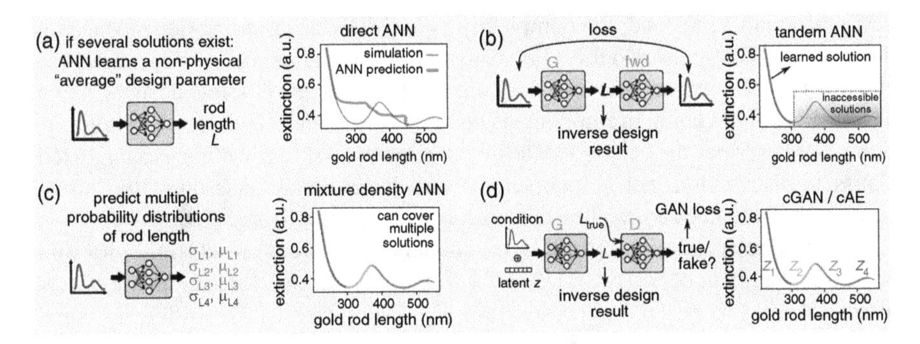

Fig. 4.3 Non-uniqueness problem in inverse design and possible solutions. (**a**) Non-uniqueness arises when a response corresponds to multiple sets of design parameters within a given precision. This "one-to-many" mapping could confuse the ANNs in reaching convergence, resulting in non-physical designs that do not coincide with any of the ground truths. (**b**) Tandem architecture. (**c**) Mixture density networks. (**d**) Conditional generative adversarial networks (cGANs) or autoencoders (cAEs). In (**c**) and (**d**), degenerated designs are coded in different colors. *G* generator, *fwd* forward simulator, *D* discriminator. (Reprinted with permission from [1] under a Creative Commons Attribution 4.0 International License (CC BY 4.0))

suffer from difficulty of convergence and fail in the attempt to suggest correct designs. A simple yet effective solution to non-uniqueness problem is cascading two ANNs in a tandem architecture [26], as shown in Fig. 4.3b. The tandem network contains a pre-trained ANN forward simulator, whose weights are fixed during the training of the other ANN for inverse design. The role of the pre-trained simulator can be understood in the following way: When the design network stands alone, the training is to minimize the error between the labeled ground truth designs and the model's prediction. With one-to-many mappings in the training data, designs corresponding to the same response lead to conflicting directions of improvement. By attaching a simulator to the output layer of the design network, the training takes a different loss function, which computes errors not between designs but responses. Because the design-to-response mapping is one-to-one, the simulator excludes all the other designs, even if they have a very close or identical response to the current prediction. By doing so, a tandem network has no trouble in reaching convergence and can make predictions correctly. The drawback, however, is that only one design of the many possible candidates can be obtained; the others are made inaccessible by the surrogate simulator.

Mixture density networks represent another logic for solving the problem (Fig. 4.3c) [27]. Instead of modeling each design parameter as a discrete value, MDNs use multiple Gaussian distributions [28], a softer treatment to evaluate the fitness of the possible designs. The probabilistic nature of MDNs ensures easy converge in the presence of degenerated solutions, which in principle are all accessible with different likelihoods. There are also a few limitations. First, the outputs of an MDN for inverse design are probability distributions of the design parameters. Usually, one still needs to sample these distributions to get the final design, and simply taking values from the most prominent peaks does not necessarily give the best performance. Second, the complexity of an MDN has a strong dependence on the choice of its multimodality. For instance, when the degree of degeneracy of solutions is unknown and the user does not want to miss any possible designs, using a large number of Gaussian components increases the models' complexity. Similarly, if the user describes the vector of N design parameters with an N-dimensional mixed Gaussian distribution, not N independent one-dimensional Gaussians, the implementation of the network will be considerably more complicated [29].

Other popular solutions to the non-uniqueness problem are cGANs [30] and conditional autoencoders (cAEs) [31, 32]. Taking the former, for example, a cGAN also contains two parts, a generator and a discriminator (or, according to some authors, a critic), as shown in Fig. 4.3d. The unique feature of the generator, compared with the standard ANNs for inverse design, is the introduction of an additional latent or noise vector at the input layer, sampled randomly from a predefined latent distribution (e.g., a Gaussian); the output designs thus follow a distribution conditioned on the input responses. The randomness of latent vectors makes the network probabilistic rather than deterministic and helps to tackle the troubles arising from degenerated solutions. Meanwhile, the discriminator serves as a "trained loss" [1], as it takes as inputs the generated designs and true designs from the training set and tries to distinguish the former from the latter. cGANs do not have the limitations of MDNs, but their training is sometimes quite sensitive to the choice of hyperparameters.

4.2 Photonic Multilayer Structures

4.2.1 Planar Multilayer Stacks

As discussed earlier, DL-based inverse design is advantageous in applications which contain a large amount of repetitive design tasks based on a common dataset. Closely related to thin-film optics and friendly for data generation, planar multilayer stacks constitute a class of photonic structures particularly suitable for testing DL algorithms [33–35]. With increasing attention and research efforts attracted, multilayer structures have the potential to be used for creating a large dataset for photonic inverse design, just like the MNIST database for image processing. Before proceeding to more sophisticated nanophotonic devices and DL algorithms, we explain for starters the usage of basic DL models in the inverse design of multilayer structures. Although some concepts have been described in the earlier discussion, a walk-through with examples further breaks them down and makes each part easier to understand.

Let us first consider the simplest scenario, a fully connected neural network being used for solving electromagnetic problems. Figure 4.4 illustrates how the network and data are configured for the forward (left) and inverse (right) problem of multilayer structures [26]. For simplicity, the ANN has only one hidden layer, although practically there are usually a few more. In this example, the multilayers

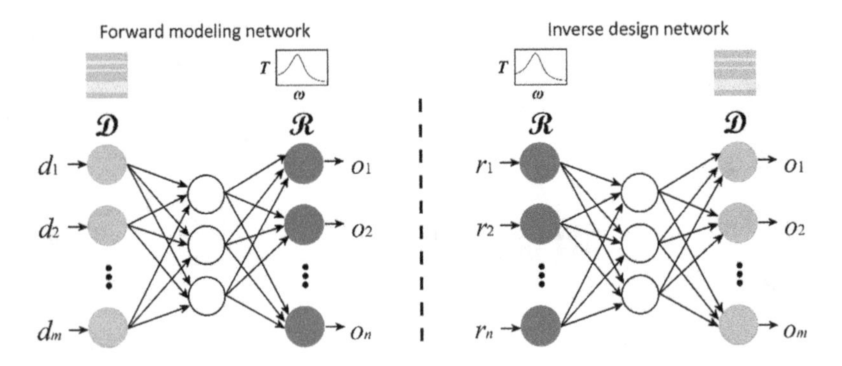

Fig. 4.4 Fully connected NNs for solving electromagnetic problems based on multilayer structures. Left: A forward modeling network (simulator) with one hidden layer. Design vector $\mathcal{D}$ comprised of thickness values for m layers is mapped to the response vector $\mathcal{R}$, which is the transmission spectrum discretized into n points in the frequency domain. Right: An inverse design network with one hidden layer. Response vector $\mathcal{R}$ is mapped to the design vector $\mathcal{D}$. Circles in blue and red denote neurons associated with design parameters and optical responses, respectively

are supposed to consist of two dielectric materials stacked alternatingly up to m layers, so the design parameters are the layer thicknesses d_1, d_2, …, d_m, forming an m-dimensional design vector $\boldsymbol{D}$. The optical response is chosen to be the transmittance at normal incidence over a frequency band, and the continuous spectra are each discretized into n points, resulting in the n-dimensional response vector $\boldsymbol{R} = (r_1, r_2, \ldots, r_n)$. In principle, with a sufficient amount of data, i.e., paired structures and their transmittance spectra, the forward modeling network and inverse design network can be trained separately, with the input and output data reversed.

In practice, the forward modeling network (or surrogate simulator) is very effective, thanks to the one-to-one mapping from design to response. Actually, except that the input design parameters sometimes need reformatting (e.g., from patterns/tensors to vectors) first through convolutional layers, fully connected NNs are always the choice for solving forward problems. However, for a naive design network, the influence of non-uniqueness problem is overwhelming. Not only will the convergence of its training be struggling, but frequent and significant inconsistency between the ground truths and predictions will occur. An example of unsuccessful design is shown in Fig. 4.5a. The target response comes from a randomly chosen 20-layer stack composed of alternating silicon dioxide (SiO_2) and silicon nitride (Si_3N_4) thin films. The response of the retrieved design, due to the model's poor efficacy, is nevertheless nothing like the ground truth. Without employing complex networks, the simplest way to resolve this issue with standard ANNs is cascading the design network and a pre-trained modeling network in a tandem architecture. While the working principle has been explained above with Fig. 4.3b, Fig. 4.5b shows how the two ANNs are actually connected [26]. Because the two networks in Fig. 4.4 have reversed structures at the input and output layers, the cascading is just straightforward with no need for additional adaption. The tandem network is proved to be effective in overcoming the non-uniqueness problem. Successful retrieval of

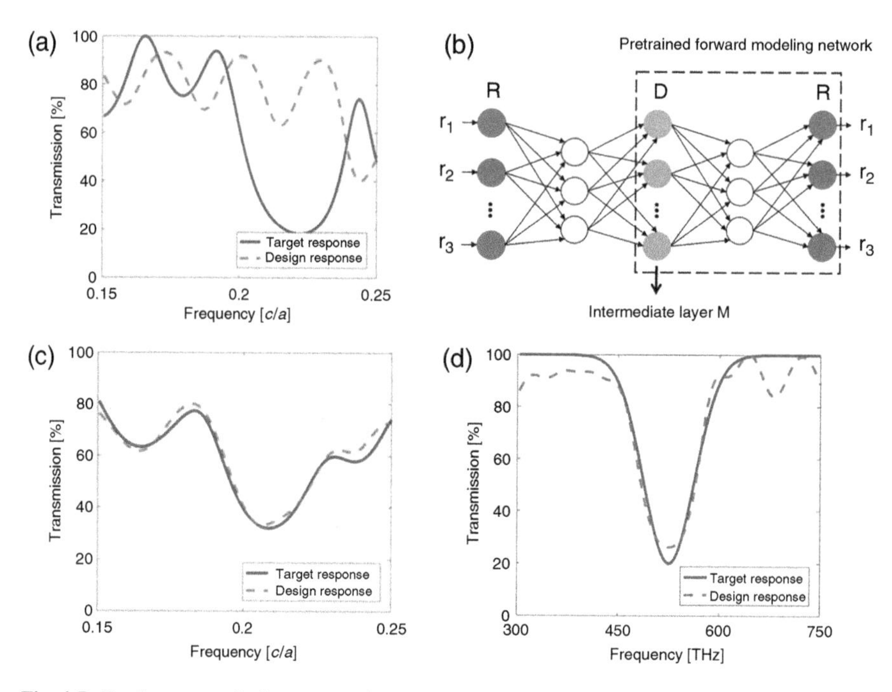

Fig. 4.5 Tandem networks for overcoming non-uniqueness problem. (**a**) Direct use of naive ANNs for inverse design leads to substantial discrepancies between target response and the response of the inversely designed structure. (**b**) A tandem network consisting of cascaded inverse design network and pre-trained forward modeling network. (**c, d**) Designs by the tandem network for a target response from the test set (**c**) and for a hypothetical target response (**d**). (Reprinted with permission from [26]. Copyright (2018) American Chemical Society)

20-layer designs for a test example and for a hypothetical response that is artificially created is presented in Fig. 4.5c, d. One thing worth mentioning again is that the retrieved design in Fig. 4.5c is not necessarily similar to the ground truth design. Tandem networks evaluate each prediction based on the fitness of its response. Therefore, the suggested design output by the intermediate layer is one of the degenerated candidates equivalent to the original one (including itself).

Next, let us move on to MDNs and see how they deal with the same design task. At this moment, we raise the requirements on device performance a little bit. In the last case, the example spectra are random, with no obvious connection to functional devices. Now, we will test whether DL algorithms can produce something useful and how well they can do the job as compared to other inverse design approaches. Figure 4.6a depicts a tandem architecture which combines an MDN for inverse design (left) and a simulator for optimization (right) [29]. Leaving the simulator apart, the MDN differs from the previous fully connected NNs mainly on the output end. While the input data, the optical responses $\boldsymbol{R}$, are still vectors of discretized reflectance spectra, the output values are not the thicknesses of all the layers but the parameters defining the probability distributions of the thickness ranges. For a 20-layer structure composed of 2 alternating materials, the output of MDN

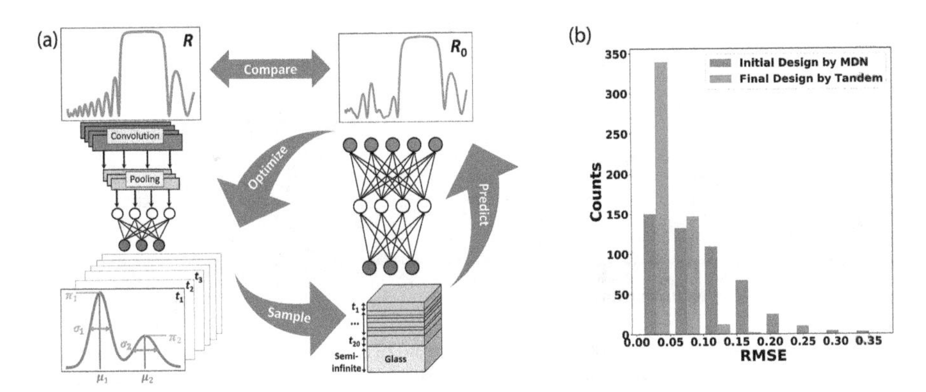

Fig. 4.6 (**a**) A mixture density network for inverse design of multilayers (left) is combined with a simulator (right) in a tandem architecture for enhanced performance. (**b**) Histogram of response RMSE before and after optimization. (Adapted from [29] with permission of Walter de Gruyter)

corresponds to 20 independent sets of mixed Gaussian distributions, which are illustrated on the lower left of Fig. 4.6a as overlaid layers. The probabilistic nature of MDNs ensures the training will not be confused by non-unique mappings in the dataset and enables access to infinite candidate designs subject to sampling of the Gaussian distributions. If an MDN is trained properly, a naive design simply taking the thickness values from the most prominent peak for each layer can already produce designs with moderate accuracy. To take full advantage of MDNs, some resampling rules have been introduced to form a combined design-optimization framework. As sketched in Fig. 4.6a, when organized in a tandem architecture, an MDN and a surrogate simulator work jointly to improve design accuracy through iterations of evaluation and resampling with the guidance of mixture probability distributions (Fig. 4.6b).

In thin-film optics, 20-layer structures are sophisticated enough to create some desirable optical responses for applications. One representative example is known as the distributed Bragg reflector (DBR) [36, 37], which features a flat high-reflection band spanning tens or hundreds of nanometers and is widely used as mirrors in optical cavities. The unusual response of DBRs is not surprising to optical scientists and engineers [38]. The underlying physics is as trivial as the constructive interference of the partial waves reflected from every interface of the multilayers, which can be achieved if the thicknesses of the layers are tuned based on their refractive indices to be one quarter of the wavelength. For a stack of alternating materials, this leads to a periodic structure. Noticing it or not, what we just explained is a process of intuition or physics-based design. But can DL algorithms do it better, or at least equally well? The answers to these questions are crucial to justify the usefulness of DL-based methods for inverse design.

Figure 4.7 compares the responses and design parameters of a target DBR and a design produced by the MDN-based tandem optimization framework. Given the fact that in the training data containing 828,000 samples (70% for training), no information about periodicity is included at all, the results are quite encouraging.

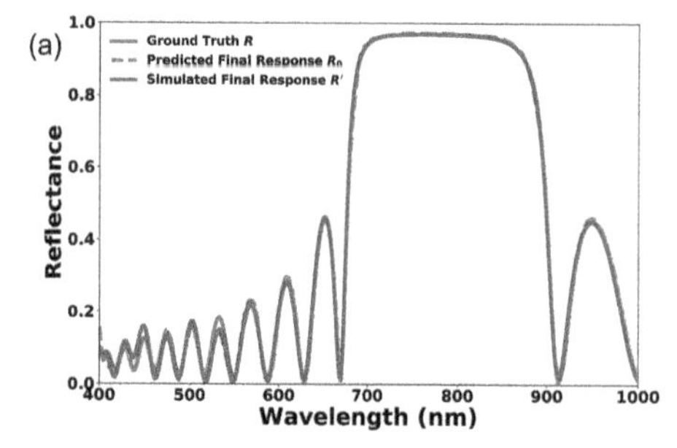

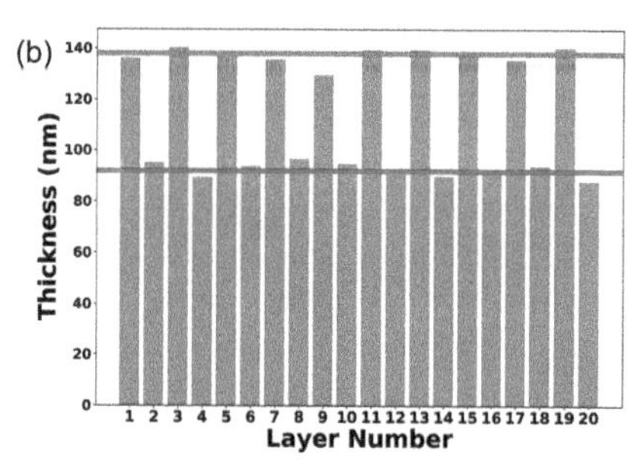

Fig. 4.7 (**a**) Comparison of the target reflection spectrum of a 20-layer DBR and the responses of an inversely designed multilayer structure. (**b**) Comparison of layer thicknesses for the target DBR (horizontal lines) and the retrieved design (columns). The two materials used are magnesium fluoride (MgF_2) and tantalum pentoxide (Ta_2O_5). (Used from [29] with permission of Walter de Gruyter)

The challenge can be further extended. Standard DBRs have a limited bandwidth, which is determined by the materials used for construction. In the frequency domain, the width of the high-reflectance zone Δf is given by [35]

$$\frac{\Delta f}{f_0} = \frac{4}{\pi}\arcsin\left(\frac{n_H - n_L}{n_H + n_L}\right), \tag{4.1}$$

where f_0 is the central frequency and n_H and n_L are the refractive indices of the high- and low-index layers, respectively. Extending the width of the high-reflectance zone is a task of practical concern. Intuition-based design strategies suggest solutions which contain two DBRs with reduced number of layers and offset central frequencies, plus a spacer in between and an optional cladding layer on the top [35]. Again, the MDN-based tandem optimization network manages to retrieve the design, as shown in Fig. 4.8. Additional tests on hypothetical broadband responses have also been demonstrated in [29].

Noticeable research efforts, using DL or DL-assisted frameworks, have been seen around the design of planar multilayer structures. From the optical side, other than ordinary responses such as transmission and reflection, inverse problems for topological states have been discussed in layered structures [39]. In parallel, attempts are being made to find better solutions of filters for thermal applications, where a broader wavelength range, a larger material library, and more layers of stacks are involved [40–43]. From the other side of data science, the training efficiency for photonic inverse design is also in study [44]. Despite the simple geometry, multilayer structures could serve as a modal system for testing DL-based design methods.

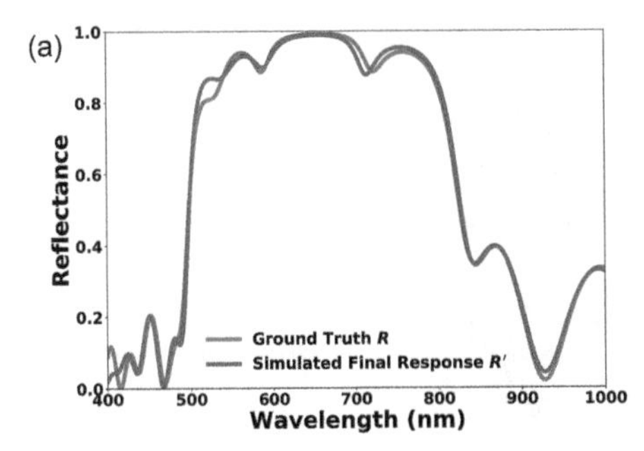

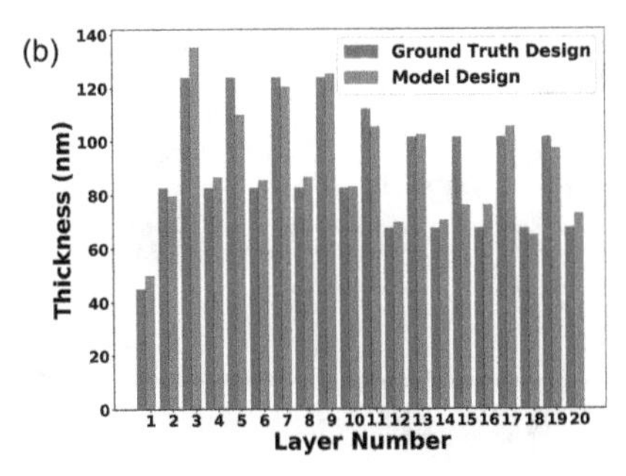

Fig. 4.8 (**a**) Comparison of the target spectrum of a 20-layer reflector with extended high-reflection zone and the response of an inversely designed multilayer structure. The target device contains two nine-layer periodic sub-stacks, with a spacing layer in between and a cladding layer on the top. (**b**) Comparison of layer thicknesses for the target reflector and the retrieved design. (Used from [29] with permission of Walter de Gruyter)

4.2.2 Multilayer Core-Shell Nanoparticles

Similar to multilayer thin films, the optical responses of core-shell particles can be analytically derived. Although many-layered nanoparticles do not have traditional applications at the industrial level as their thin-film cousins do, the interesting properties they possess are inspiring for new opportunities in, e.g., solar cells, radiative cooling [45], and optical cloaking [46]. The inverse design of multilayer core-shell nanoparticles can be done with no doubt using the algorithms introduced above, but here we would like to discuss another possible way in which DL models can be used for design purposes.

As a schematic, Fig. 4.9a illustrates the structure of a three-layer nanoparticle and a fully connected NN. With the previous preparation, it is easy to recognize this NN functions as a surrogate simulator for computing the optical responses of the particles. In some early attempts of inverse design in electromagnetics, the usage of NNs is limited to solving the forward problem to accelerate the optimization process [20], during which the most time-consuming part is the simulations performed by electromagnetic solvers such as FDTD, FEM, FIT, etc. To some extent, the role of the pre-trained simulator in tandem networks is the same. In addition to this hybrid strategy and the inverse design NNs, interestingly, a surrogate simulator can be operated reversely as an optimizer [47], providing the third way that a NN can be used for inverse design. The mechanism of this working mode is based on backpropagation. During the training of the simulator, weights are iteratively adjusted through backpropagation to reduce the error between the target responses and predicted responses. For inverse design, the error still measures the distance to the target responses, but the weights are fixed. Therefore, the variables being updated are the design parameters at the input layer, driven essentially by a gradient-descent search. Figure 4.9b compares the design results for an eight-layer nanoparticle made of alternating shells of titanium dioxide (TiO_2) and SiO_2. Whereas the

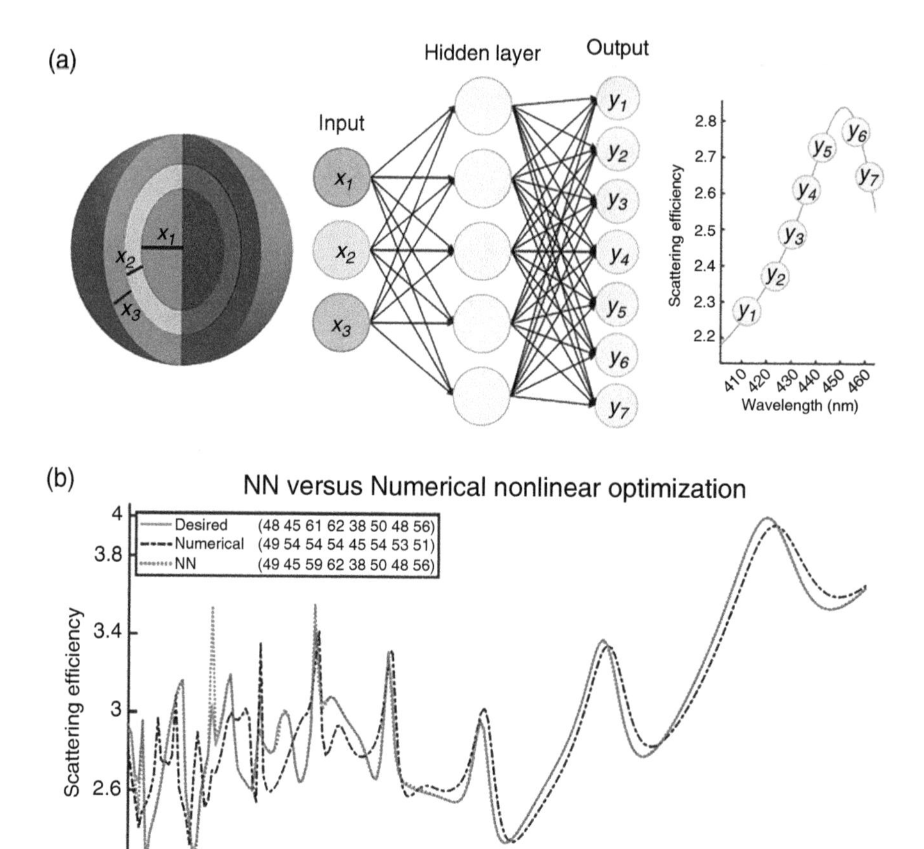

Fig. 4.9 (**a**) A fully connected NN as a simulator for the scattering of multilayer core-shell nanoparticles. (**b**) Gradient-based inverse design for an eight-layer core-shell nanoparticle (red dotted line), in comparison to the target spectrum (blue line) and a design based on interior-point optimization (black dashed line). Dimensions of the eight layers (unit: nm) by different methods are given in the legend. (Reprinted from [47] with permission. Copyright (2018) The Authors, some rights reserved; exclusive licensee American Association for the Advancement of Science. Distributed under a Creative Commons Attribution-NonCommercial License 4.0 (CC BY-NC 4.0))

gradient-descent method appears very accurate, another nonlinear optimization based on interior-point methods shows noticeable inconsistency. A noteworthy limitation of this gradient-descent method is the final result highly relies on the initial guess: it can be one of the degenerated solutions if non-uniqueness is present, or it can get trapped at an undesired local optimum and does not suggest any correct design. When the latter happens, the design needs to start over with a new initial guess [2].

So far, our examples only contain geometric parameters (film/shell thicknesses) in the design vector. It is natural to wonder whether and how other aspects of the inverse problem, as named in Fig. 4.1, can be dealt with by DL models. In a study of MDNs for multilayer design, the angle of incidence has been added to the design vector [27]. Later we will also see an example of metasurface where the polarization of the incidence is taken as a design variable [49]. Figure 4.10 shows another design task for multilayer core-shell particles [48]. A particularly interesting part is, in addition to the shell thicknesses, the materials of the core and shells are to be selected from a group of candidates as well. The complete network has a structure similar to the tandem architecture (Fig. 4.10b), where an intermediate layer of design parameters connects the inverse design module and the forward modeling module, but neither of the modules is pre-trained. In the design vector, the labels of material types are treated with no difference from shell thicknesses despite their

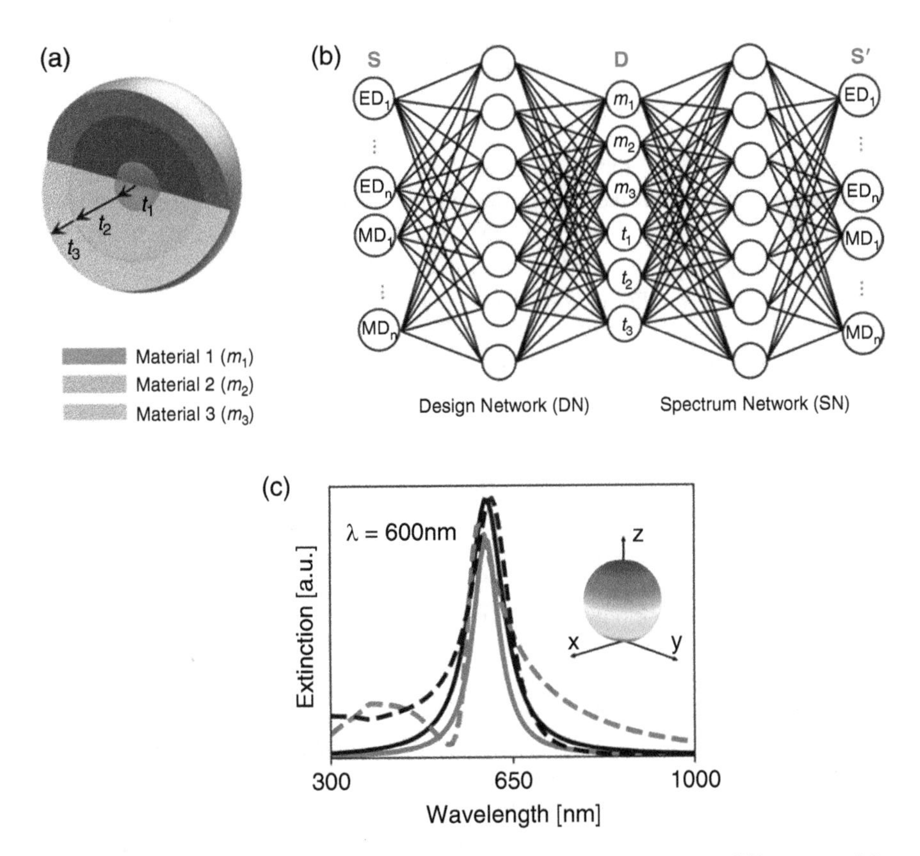

Fig. 4.10 (**a**) Schematic of a three-layer core-shell nanoparticle made of three different materials. (**b**) An inverse design network in which geometric parameters (t_1–t_3) and material types (m_1–m_3) are simultaneously encoded in the design vector. (**c**) Target spectra (solid curves) and predicted spectra (dashed curves) for ED (red) and MD (black). Inset: Far-field pattern showing directional scattering at the wavelength of 600 nm. (Reprinted with permission from [48]. Copyright (2019) American Chemical Society)

different natures. Nonetheless, the loss function does need adaptions to have a properly weighted evaluation of errors from the predicted responses, shell thicknesses, and identification of materials. The training of this network turns out to be sensitive to some coefficients in the loss function, and it is not clear yet whether this can be avoided by using a tandem network. Figure 4.10c shows the design of a three-layer nanoparticle with spectrally overlapped ED and MD under the excitation of a linearly polarized plane wave. Although the desired spectra are hypothetical, the design captures the major features of interest. With close magnitudes and phases near the wavelength of 600 nm, the ED and MD of the nanoparticle fulfill the first Kerker condition, giving rise to directional scattering in the forward direction.

4.3 Metasurfaces

The diversity of the structures and functionalities of metasurfaces implies more powerful tools to be used for inverse design. When the building blocks are resonant particles in regular shapes, the metasurface design can be described by a set of parameters [50–54], and the complexity, if estimated merely by the dimension of the design vector, is not necessarily higher than that of a thin-film structure with many layers. In this case, the design can be performed in a fashion like what have been discussed above, where networks produce deterministic outputs.

The recent interests in applying DL to photonic inverse design were probably inspired, at least partially, by an attempt to design metasurface-like plasmonic nanostructures via DL (Fig. 4.11) [49] (the early version was posted on arXiv in February 2017). The function of the target nanostructures is not phase-related like beam shaping or steering, but conceptually closer to that of thin films to produce desired spectral features in transmission, except that the responses of nanostructures can be polarization-dependent. Despite the different physical mechanisms leading to the spectral features (interference for thin-film structures versus resonances for nanostructures), data-driven design methods do not treat them very differently. Figure 4.11a shows the architecture of the network comprising cascaded inverse and direct networks. At the input layer, two sets of spectra for horizontal and vertical polarizations and material types are fed into three parallel group layers, followed by fully connected layers to predict the designs. Eight geometric parameters are used to describe the nanostructure in the shape of letter H. And again, the subsequent simulator is a fully connected NN, making predictions about the transmission based on received geometry, material types, and indicators of the polarization. The resonance behaviors of H-shaped particles, compared with those introduced in Chap. 2, are markedly more difficult to comprehend for a human. One with knowledge of nanophotonics at some level might be able to deduce some of the possible ingredients like detuned ED modes in the asymmetric vertical arms and MD modes from the SRR-like super- and substructures. However, these are not informative enough to guide the design effectively, which conventionally is accomplished with a

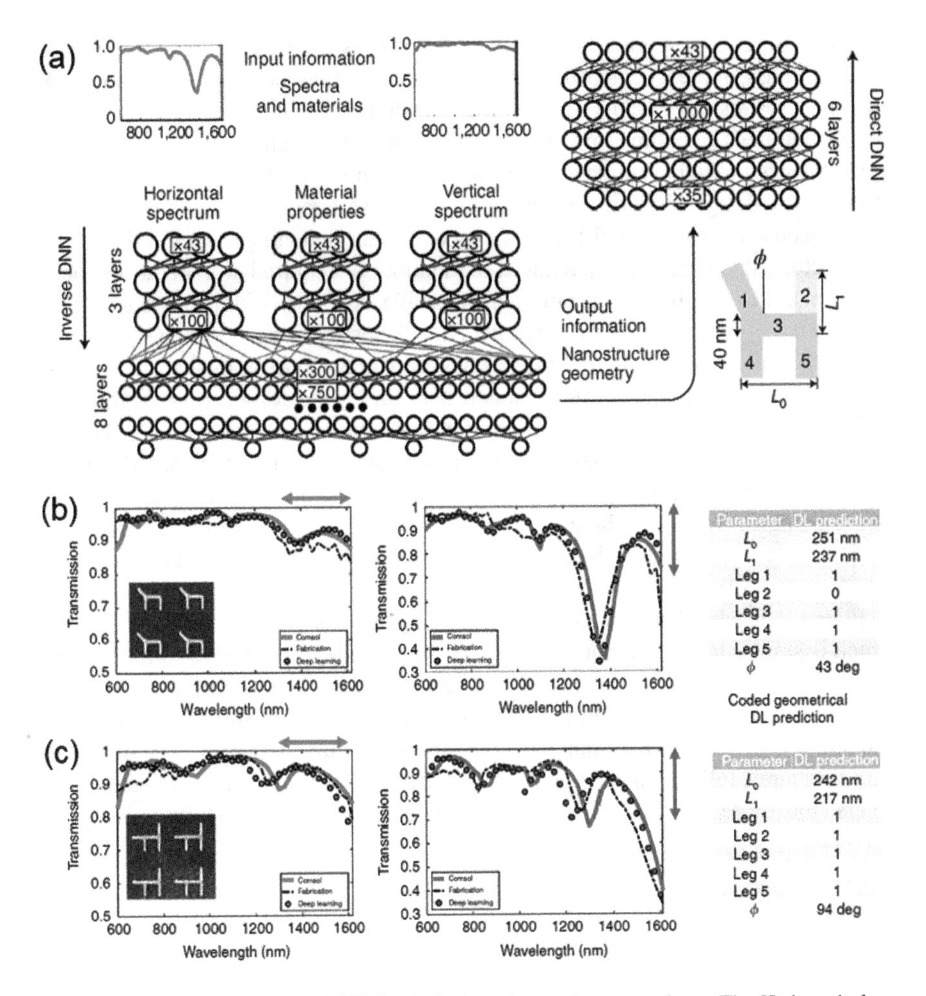

Fig. 4.11 (**a**) A fully connected NN for designing plasmonic metasurfaces. The H-shaped plasmonic unit cell is characterized by eight parameters, of which five denote the existence of the legs by a binary number (0/1) and three define the leg lengths and the tilting angle of leg 1, respectively. (**b**, **c**) Results of inverse design for spectra of two different samples shown in the insets. Insets: SEM micrographs of fabricated samples. (Reprinted from [3] (for an adapted version of panel (a)) with permission of Springer Nature and from [49] (for all panels) under a Creative Commons Attribution 4.0 International License (CC BY 4.0))

thorough parameter sweep and, if needed, assisted by gradient-descent or other optimizations.

Unlike most demonstrations that are purely numerical, the work by Malkiel et al. further tested the model with measured spectra of fabricated samples. Results for two representative devices are shown in Fig. 4.11b, c. Considering that each spectrum is only discretized into 43 data points, the accuracy of the retrieved designs is reasonably good.

Going beyond structures in regular shapes proves an effective way to achieve better or even unprecedented device performance. Many successful examples in photonic crystals and circuits have been demonstrated by using advanced optimization techniques [7], ending up with very irregular and nonintuitive design patterns. Metasurfaces provide a suitable stage for DL to join the competition. For this type of tasks, the design layouts are usually processed as image data by convolutional and transposed convolutional layers. Compared with flattening discretized images into vectors, this significantly reduces the number of required weights in the network, making the training much more manageable, and allows for better learning of high-level features in the design layouts. Nonetheless, given the vast design space and non-uniqueness problem, simply stacking (transposed) convolutional layers into a tandem architecture is often inadequate to survey the design space thoroughly and efficiently. Instead, generative models with stochastic components (e.g., GANs and VAEs) are superior options, which offer dramatically enhanced capabilities of learning rich distributions in image data and suggesting designs that contain complex topological features. There is, of course, no restriction on applying generative models to devices described by discrete parameters. Here, we just shift our focus to free-form metasurfaces.

Figure 4.12 sketches the architecture of a cGAN for photonic inverse design, which is similar to the structure in Fig. 4.3d but includes a pre-trained simulator to evaluate the optical responses of the generated patterns. The loss function used to train the generator is therefore modified to account for errors from both the critic and simulator. Because the inputs and outputs of the three networks have different data structures, fully connected layers (for spectra and true/false) and convolutional layers (for discretized images) are used at the ends, with hidden (fully connected,

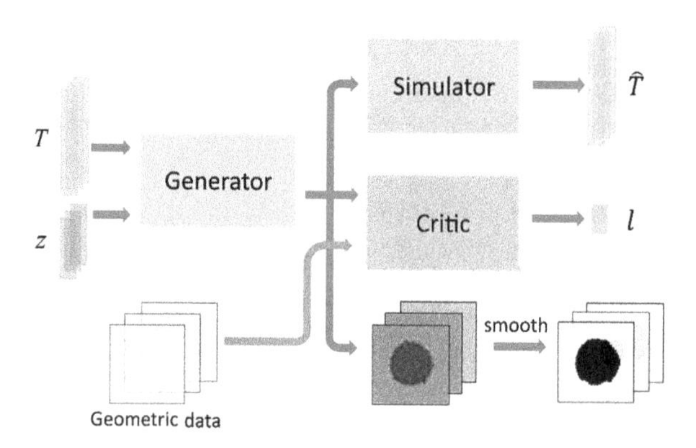

Fig. 4.12 Structure of a cGAN for photonic inverse design. A pre-trained simulator is added to the generator-critic system. During training, weights in the generator are updated through backpropagation of the losses defined by the simulator and critic, while the critic itself uses the loss following the Wasserstein GAN. (Reprinted with permission from [30]. Copyright (2018) American Chemical Society)

(transposed) convolutional, pooling) layers properly configured and ordered between them. Figure 4.13a compares a few representative patterns from the test dataset and the corresponding patterns produced by the cGAN. Noticeably, the generated patterns highly resemble the original ones, not only in shape but also in their polarization-resolved responses (Fig. 4.13b, c). More interestingly, when the model

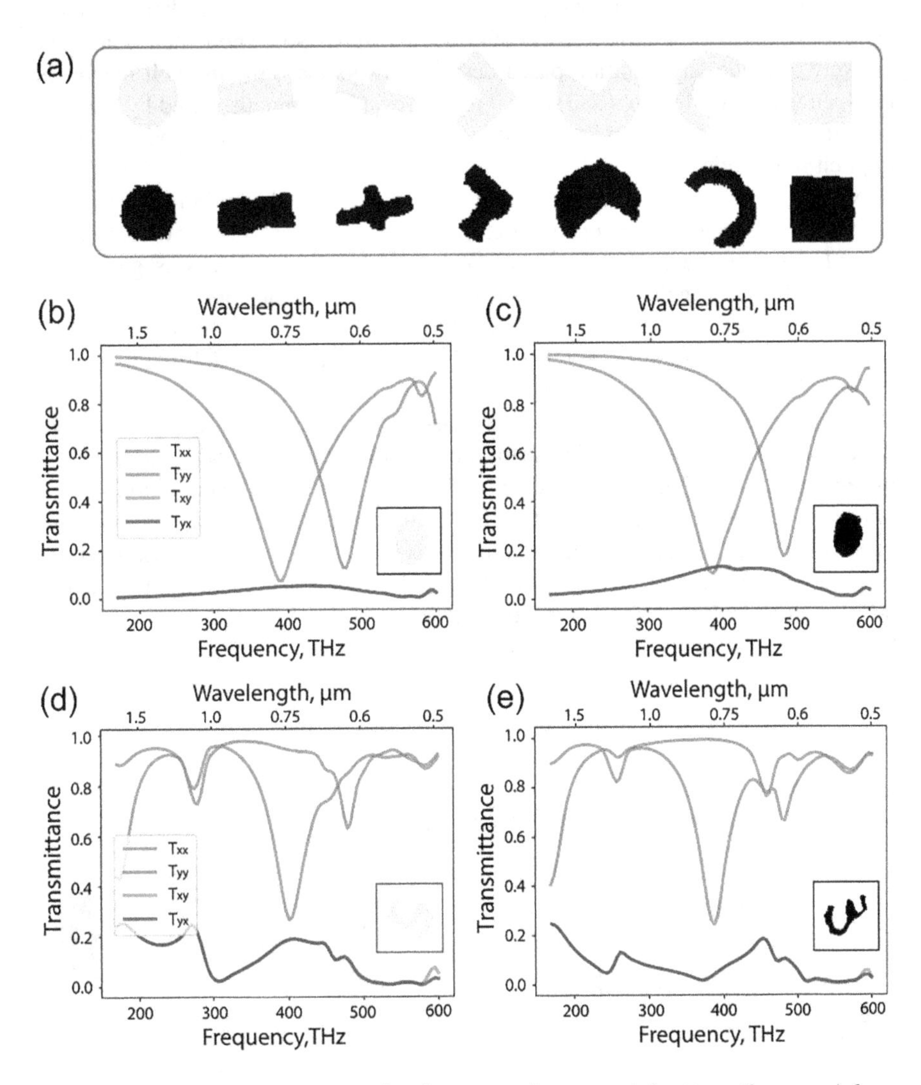

Fig. 4.13 (**a**) Test patterns (upper row) and corresponding generated patterns (lower row) for a cGAN trained on a predesigned class of geometry data. (**b, c**) Polarization-resolved transmittance of a test sample (**b**) and of the generated pattern (**c**). Results are achieved when the critic is trained on the data of the elliptical class. (**d, e**) Same as (**b, c**) but for a model trained on a modified MNIST database. The network generates a modified "3" to best replicate the spectra for pattern "5," which was removed intentionally from the training data. (Reprinted with permission from [30]. Copyright (2018) American Chemical Society)

is tested with patterns from a class of geometric data not seen during training, it can still suggest alternatives that have similar optical responses. A demonstration is presented in Fig. 4.13d, e. When the cGAN was training on a modified MNIST database, the class of digit 5 was intentionally excluded. However, upon receiving the spectra of a metasurface composed of an array of unit cells in the shape of 5, the network retrieves a design with digit 3. From the human perspective, there are some similarities in the geometric features of these two digits, which nonetheless are very unlikely to guide any intuition-based design. cGANs learn the implicit mappings between geometric features and optical responses. Not being distracted by the concepts of "digits" or "letters," they derive the design layout simply based on the implicit information.

VAEs can be used in a similar fashion. One example is shown in Fig. 4.14 [31]. Different from standard autoencoders employed in an earlier work for dimensionality reduction to alleviate the non-uniqueness problem [52], VAEs as generative models solve it by inclusion of a latent space between the encoder and decoder. In essence, the architecture of the conditional VAE in Fig. 4.14 can be understood by analogy with the case of cGAN in Fig. 4.12. The generation model functioning as the decoder takes optical responses and sampled latent variables as inputs to reconstruct the patterns encoded in the latent space by the recognition model. Although there is no discriminator to trick with, the reconstruction error needs to be minimized during training the whole ensemble, by which the recognition model (encoder) gets optimized. Flexibility exists in the way of involving optical response data. Other than using pre-trained simulators as in the previous examples, it is tractable to integrate a prediction model (simulator) in the system and train the entire network in an end-to-end manner. Other strategies without using surrogate simulators [55] or having an extra adversarial block [56] have also been used in recent works. One special feature of VAEs for inverse design is that the structure of the

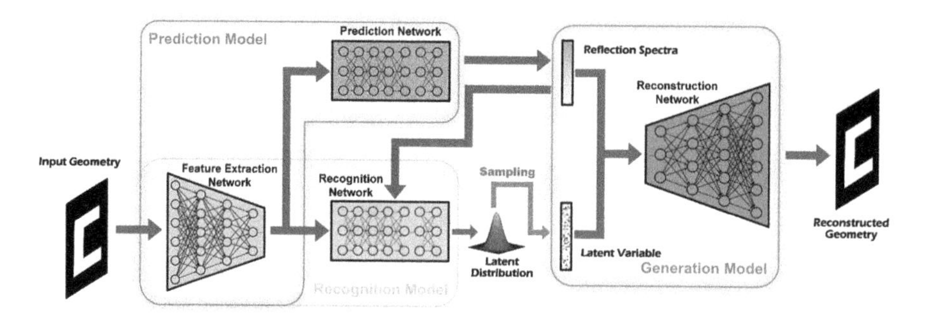

Fig. 4.14 Architecture of a conditional VAE for metasurface design. The recognition model (green box) encodes the input patterns of meta-atoms with their optical responses generated by the prediction model (blue box) into a multivariate latent distribution, which is then sampled to provide latent variables to the decoding part, i.e., a generation model (orange box) conditioned on the predicted spectra. (Reprinted with permission from [31]. Copyright (2019) Wiley)

latent space, where representations of the device patterns are encoded, can be visualized using dimensionality reduction techniques such as t-SNE. This could provide possibly interpretable insights to understand the generated designs [31, 57].

4.4 Photonic Structures with Chiroptical Responses

In Chap. 2, we introduced metasurfaces and emphasized their 2D nature, which is manifested by subwavelength-sized and spaced optical antennas on a single interface. As research on this topic advances, novel phenomena and applications based on devices consisting of more than one layers of antenna arrays have been demonstrated. Albeit the resulting increase of device thickness may not change the dominant role of abrupt phase jumps and critical coupling over the propagation effect, it sometimes blurs the boundary between metasurfaces and metamaterials. For our context, neither sticking to a certain preference of classification (and use of terminology) nor forcing such devices into one of the two categories case by case is necessary. We simply treat these “layered metasurfaces” or “ultrathin metamaterials” as photonic structures and, in the following, discuss their connections to chiroptical responses.

Chirality is intrinsically a 3D property. A truly planar structure is achiral, because it will be superimposable on its (in-plane) mirror image if the plane it lies on is flipped over. Practically, nevertheless, planar chiral structures are made possible for several reasons [59]. To name a few, fabrication imperfections always render planar elements asymmetric in the thickness direction, the presence of a substrate breaks the out-of-plane mirror symmetry, and extrinsic chirality can be induced by oblique incidence [60]. Despite all these enabling factors, 3D nanostructures like helices with intrinsic chirality are generally believed to have stronger chiroptical responses, restricted only by the fabrication complexity. As discussed in Chap. 2, stacking layers of planar resonant building blocks provides an alternative pathway to creating strong chiroptical responses while retaining the ease of fabrication. The key to success is the judicious design of the structures, which needs to deal with the resonances of unit cells and their interactions in a more sophisticated way.

Figure 4.15 summarizes an interest attempt made by Ma et al. [58]. The chiral metamaterial shown in panel (a) comprises two layers of periodically arranged SRRs over a backplane. With the opening of SRRs in the bottom layer parallel to the lattice, chirality is introduced by twisting the orientation of SRRs in the top layer, giving rise to selective reflection of CPL. At a first glance, the design task seems to have nothing special. Because the device performance is characterized by the contrast between the reflection amplitudes for LCP and RCP incidence, one may expect that a tandem network or its equivalent would suffice. However, this idea is not well supported by Ma’s experiments. Using the primary network in Fig. 4.15b, where the polarization-resolved reflection spectra are processed by three parallel channels in the forward path, the simulator proves not effective in capturing resonant features

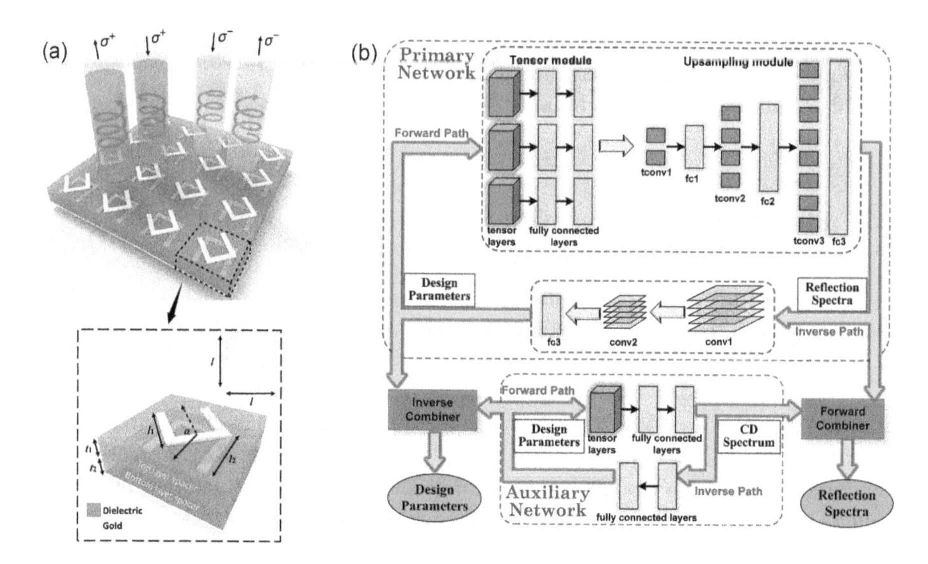

Fig. 4.15 (**a**) Schematic of a chiral reflect-array metasurface/metamaterial, which selectively reflects CPL of one handedness. Inset: Structure of a unit cell defined by five geometric parameters. Two layers of gold SRRs with a twisting angle are stacked above a backplane, separated by dielectric spacers. (**b**) Structure of an ANN for designing photonic devices with chiral responses. The model combines a primary network (PN) and an auxiliary network (AN), which are stacked with the connection of two combiners. PN and AN operate independently to deal with the same design task based on polarization-resolved reflection and circular dichroism (CD) spectra, respectively, and the results are mixed for further refinement by the two combiners. *fc* fully connected layer, *(t)conv* (transposed) convolutional layer. (Panel (b) is reprinted from [58] with permission. Copyright (2018) American Chemical Society)

(see Fig. 4.16a). To improve the predictive capability, an auxiliary network is trained to directly model the differential absorption of LCP and RCP (i.e., circular dichroism (CD)) derived from the reflection spectra and is integrated with the primary network using an ensemble method [61]. The auxiliary network is also in the tandem architecture, and as shown in Fig. 4.16b, it handles CD spectra with high precision, providing threshold information to the combiners for refining the results produced by the primary network. The combined network in Fig. 4.15b receives improved prediction accuracy, leading to not only on-demand inverse design of a certain type of chiral metamaterials (Fig. 4.16c, d) but also the discovery of some counterintuitive conditions that maximize the chiroptical responses.

It should be pointed out that the performance of NNs can be task specific. In the present case, although the discrepancy in the primary network's output is attributed to the rarity of data points near the sharp resonances, a different model may totally resolve it without the help of an auxiliary network. For a given task, one usually begins by evaluating a few models and chooses the best one to fine-tune, whereas ensemble methods such as stacked generalization introduced in [62] are useful strategies when single networks alone do not perform well enough.

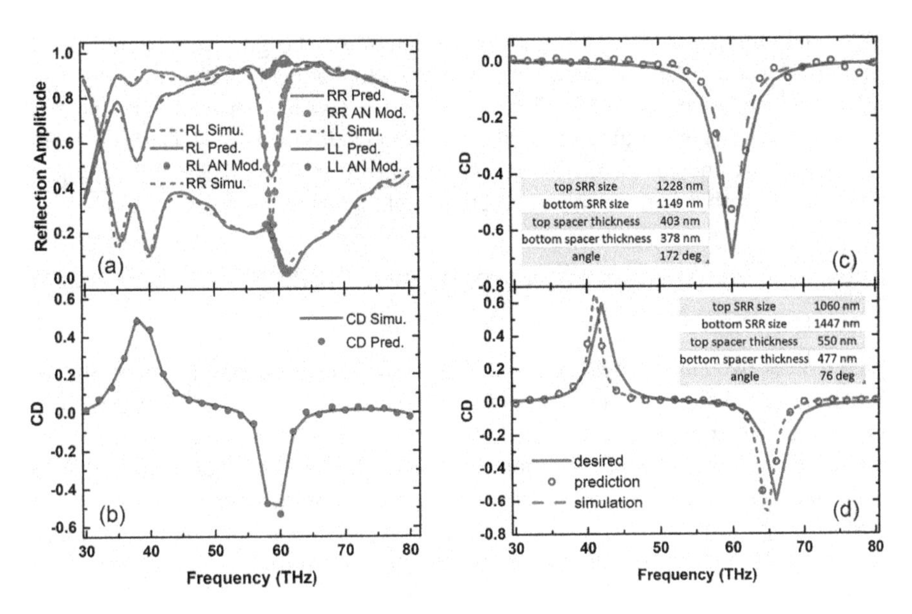

Fig. 4.16 Forward prediction (**a, b**) and inverse design (**c, d**) performance of the model in Fig. 4.15. In (**a**), results produced by the PN without AN (solid curves) are included for comparison. RL/RR/LL describes polarization-resolved reflection. The first letter denotes the handedness of the reflected light, and the second letter indicates that of the incidence. For cross-polarization conversion, RL and LR are equal in amplitude. In (**b**), the prediction is made by the AN. (Reprinted from [58] with permission. Copyright (2018) American Chemical Society)

4.5 Deep-Learning-Assisted Optimization for Inverse Design

While DL and optimization techniques have been compared extensively in the recent literature as two classes of approaches for inverse design, there is by no means any rule prohibiting their integration in a design framework. In fact, as the strengths and weaknesses of DL and conventional optimization are complementary on some measures (e.g., accuracy, computational cost, capability of searching global optima, etc.), it is rather interesting to work on the borderline where methods from both sides are combined so that even better performance can be achieved. In this section, we discuss two representative examples that showcase the joint efforts by DL and conventional optimization or, in particular, TO. The use of other optimization methods like generic algorithms can be understood similarly [63].

TO has proved a very useful tool for generating high-performance designs of photonic devices [7, 8]. It typically begins with a random distribution of dielectric functions, which is discretized with each pixel taking either the refractive index of a candidate material or that of air. This initial design is optimized through iterative evaluations of device performance (i.e., solving the forward problem), computation of gradients for directions of improvement, and updates of the device layout. Given the large number of design variables in the problem, determining the gradients with the brute force is computationally very expensive. This challenge, nevertheless, can be solved by using the adjoint method [8, 64–66], which employs only one additional simulation in every iteration to determine the gradients, significantly reducing the computational cost. Without turning to the theory or formulation, here we outline the procedure briefly. For a given design task, an objective function F is defined based on some figure of merit, such as the field intensity at a specific location or the efficiency of diffraction in a certain direction. The gradient of the objective function with respect to the local refractive index $n(\mathbf{r})$ can be expressed by

$$g = \frac{\partial F}{\partial n(\mathbf{r})} \propto \mathrm{Re}\left\{\mathbf{E}_{aj}(\mathbf{r}) \cdot \mathbf{E}_{og}(\mathbf{r})\right\} \tag{4.2}$$

that involves two electric fields $\mathbf{E}_{og}$ and $\mathbf{E}_{aj}$. The first field $\mathbf{E}_{og}$ is simply the solution of a direct simulation which evaluates the current design as solving a forward problem, and the other one, $\mathbf{E}_{aj}$, is obtained from an "adjoint" simulation, where the excitation is changed based on the function of the target device. For instance, if the device is to maximize the electric field intensity at a point $\mathbf{r}_0$, in the adjoint simulation, the current design is excited by an electric dipole at $\mathbf{r}_0$ [66]. With the gradient information known from these two simulations per iteration, the refractive index profile $n(\mathbf{r})$ can be adjusted constantly until it converges to a design with desired performance but no guarantee of global optimality [7]. When additional design suggestions are required, the whole procedure needs to start over with different initial guesses.

In general, for an inverse design task, although there is no provable guarantee of a globally optimal solution, applying global search techniques may result in a better chance of finding high-performance designs. One possible realization of such strategies using DL was recently proposed by Jiang et al. for optimization of metasurfaces [67]. Termed global topology optimization networks (GLOnets), the method combines adjoint-based computation with generative models. As depicted in Fig. 4.17a, the conditional GLOnet does not have a discriminator/critic or surrogate simulator like cGANs. Instead, a generator is connected to a module where physics-based gradients are calculated based on the adjoint method using an electromagnetic solver. In other words, the network learns directly through simulations. This lifts the requirement of a training dataset. It is then important to see how a GLOnet differs from adjoint-based TO. In the current example, the target metasurface consisting of silicon gratings deflects normally incident light of wavelength λ to the +1 diffraction order at angle θ; see Fig. 4.17a for schematics of the direct and adjoint problems. When the device is discretized into m pixels, during the course of TO from iteration k to $k + 1$, the refractive index of the i-th element is updated following

$$n_i^{k+1} = n_i^k + \gamma g = n_i^k + \gamma \frac{\partial E}{\partial n_i}, \tag{4.3}$$

where E is the diffraction efficiency for the wavelength and angle pair and γ is a step parameter subject to tuning. This process applies to one device and takes place in a local region of the design space, which is determined by the initial layout of the dielectric functions. In contrast, promised by the nature of the generator, a conditional GLOnet optimizes a distribution of devices. During each iteration of training, the updates of weights take the efficiency gradient averaged over a batch of devices:

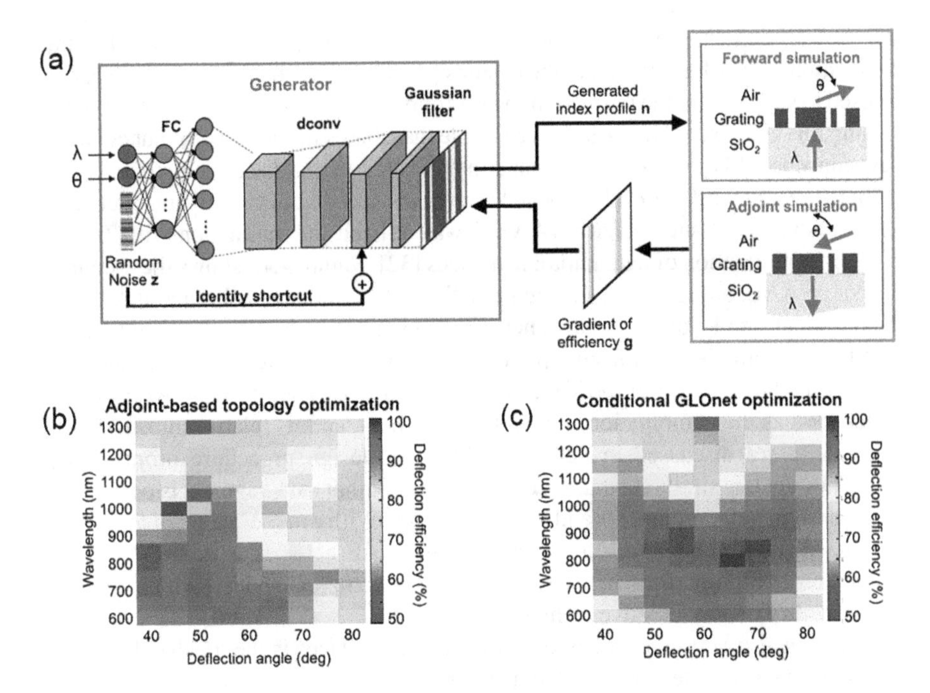

Fig. 4.17 (**a**) Schematic of the conditional GLOnet optimizer, which combines a generator (left, blue box) and an adjoint-based process (right, orange box) providing physics-driven gradients. (**b**, **c**) Performance comparison of adjoint-based topology optimization (**b**) and condition GLOnet optimization (**c**). Plots for both methods are based on devices of the highest efficiency. For each cell that corresponds to a certain wavelength and angle combination, its value is taken from the best device out of 500 optimizations (each runs 200 iterations of forward/adjoint simulations) or 500 designs generated by the network. (Reprinted with permission from [67]. Copyright (2019) American Chemical Society)

$$w_i^{k+1} = w_i^k + \eta \frac{\overline{\partial E}}{\partial n_i}. \tag{4.4}$$

Compared with optimizing individual devices from randomly generated initial layouts with TO, conditional GLOnets allow for crosstalk between device instances seen during the whole training process. Moreover, mediated by the mapping from input noises z, the design space is sampled globally, which turns out to be effective in searching high-performance regions [67]. Figure 4.17b, c compares the best designs by TO and the GLOnet. While the latter produces better results for more wavelength and angle combinations, it consumes comparable or even fewer computational efforts: The total number of (direct and adjoint) simulations used for training the GLOnet is 2.5 million, which is about the amount for adjoint-based TO if 50 devices are optimized for each wavelength and angle pair.

Another route via DL to enhancing optimization is to have the networks learn from optimized designs instead of randomly generated ones. Several successful attempts have been reported in this direction, such as training cGANs on a small set of topology-optimized high-performance designs of metagratings [68]. Starting with structures already in some favorable regions of the design space helps to alleviate the data requirement of training, and the designs produced later by the trained ANN can be used to supplement the initial training set following various strategies [69]. Before closing this chapter, we introduce an example that shares the same spirit but contains new concepts of DL and of metasurfaces not covered in our discussion so far.

In the recent work by Kudyshev and coworkers, a different type of autoencoders, termed adversarial autoencoders (AAEs), was utilized in conjunction with TO for efficient optimization of nanophotonic devices [32]. Figure 4.18 shows the architecture of an AAE, along with the design pipeline. The major new idea about AAEs is the introduction of a discriminative network, which is attached to a VAE and judges whether a sample is taken from a predefined model distribution (positive samples) or from the latent space of the VAE (negative samples). Therefore, an AAE can be understood as the combination of a VAE and a GAN, and its relationships to VAEs and GANs are discussed in detail in [70]. The design procedure proposed by Kudyshev et al. is similar to that used in [68], except that data augment is performed based on symmetry arguments and prior to training. With a moderate set of topology-optimized structures for the target function, the AAE network is trained for rapid generation of high-performance and topologically complex devices, which are then refined with TO or alternative techniques [32].

The optical function to be achieved by the metasurfaces is the idealized thermal emission. As illustrated in Fig. 4.19a, for photovoltaic cells that have a working band ranging from 0.5 to 1.7 μm, the radiation by an ordinary heater overlaps only a small portion of the band. To improve the efficiency of electrical power generation, the device needs to be optimized towards an ideal emitter whose emission spectrum fully coincides with the cell's working band. According to Kirchhoff's

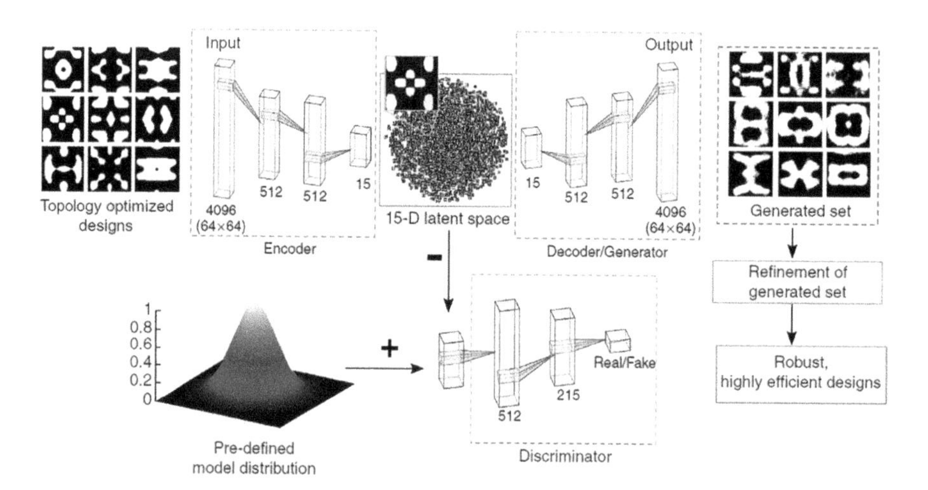

Fig. 4.18 Schematic of AAE-assisted TO for metasurface design. An AAE contains a VAE (top row) and an adversarial network (i.e., the discriminator) that guides the latent space to match a predefined model distribution (lower left). The AAE is trained on a training set of high-performance designs generated by TO, and upon completion of training, structures produced by the AAE are further refined. (Reprinted from [32] with permission of AIP Publishing)

law of thermal radiation, the emissivity and absorptivity of an object are equal. Hence, developing an ideal emitter is practically equivalent to designing a perfect absorber operating over the same wavelength range. The mechanism and design principle of metamaterial and metasurface absorbers have been documented by several nice reviews [71–73], but practically, it is still not a trivial task to find the exact structure for total absorption over a specific band with material constraints. The search is usually assisted by optimization techniques. For the present problem, the base structure for optimization is shown in Fig. 4.19b, where the top layer takes a simple form of a continuous thin film without patterning, and titanium nitride is used as an alternative plasmonic material due to its superior resistance to high temperatures [74]. The performance of the devices generated by direct TO and AAE-assisted TO is compared in Fig. 4.19c, d. Noticeable improvement brought by the use of AAE can be observed in both statistics and for the best designs.

As a relatively new tool for photonic inverse design, DL has shown promising results in some applications and by certain measures. Despite the exciting ongoing

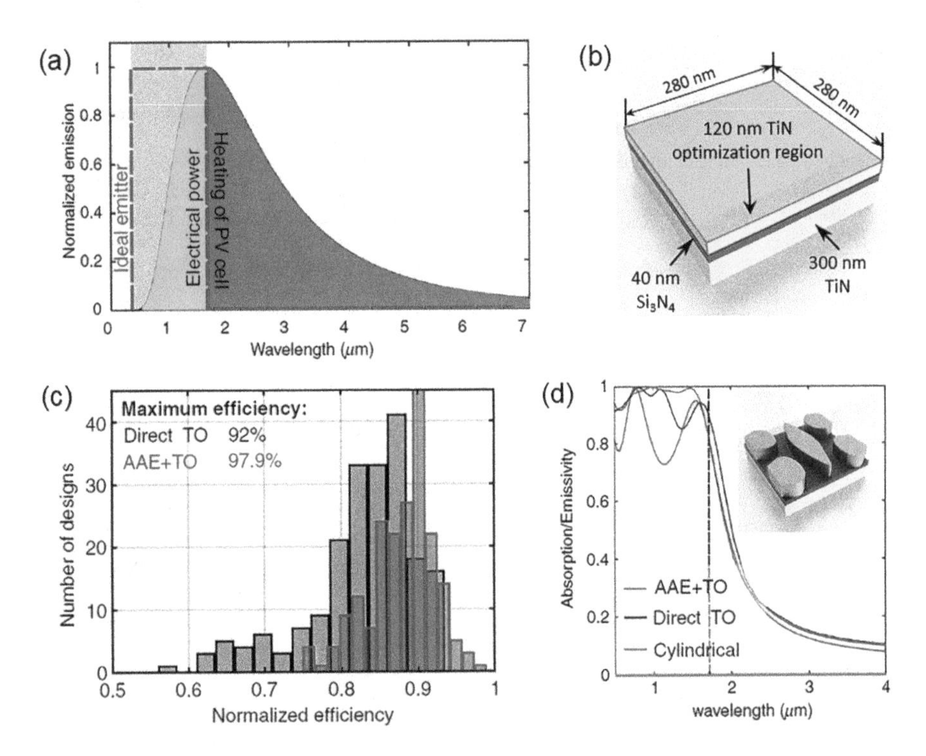

Fig. 4.19 (**a**) Radiation of a thermal emitter corresponding to the blackbody emission spectrum at 1800 °C (black curve), in comparison to the radiation of an ideal emitter (dashed blue curve) that perfectly overlaps the working band of a photovoltaic cell (gray region). (**b**) The base structure of a unit cell of the metasurface thermal emitter. Only the top layer is optimized. (**c**) Comparison of the statistics of 200 designs produced by TO and by AAE-assisted TO. (**d**) Comparison of the absorption/emissivity of the best designs (including a circular disk array) obtained by using different methods. Inset shows the AAE-generated structure after TO refinement. (Reprinted from [32] with permission of AIP Publishing)

progress, there remain a lot of open questions at different levels, ranging from, e.g., what other applications involving physics with high degrees of abstraction could be studied, how a suitable model can be determined for a given design task, whether establishing databases of photonic devices (like the MNIST database) is necessary and possible for testing design methods, to how new paradigms of inverse design can be developed based on existing DL and optimization techniques [75]. The exploration has just begun.

References

1. Wiecha, P.R., et al.: Deep learning in nano-photonics: inverse design and beyond. Photonics Res. **9**(5), B182–B200 (2021)
2. Jiang, J., Chen, M., Fan, J.A.: Deep neural networks for the evaluation and design of photonic devices. Nat. Rev. Mater. **6**(8), 679–700 (2021)
3. Ma, W., et al.: Deep learning for the design of photonic structures. Nat. Photonics. **15**(2), 77–90 (2021)
4. Liu, Z., et al.: Tackling photonic inverse design with machine learning. Adv. Sci. **8**(5), 2002923 (2021)
5. Khatib, O., et al.: Deep learning the electromagnetic properties of metamaterials—a comprehensive review. Adv. Funct. Mater. **31**(31), 2101748 (2021)
6. Chen, M.K., et al.: Artificial intelligence in meta-optics. Chem. Rev. **122**, 15356 (2022)
7. Molesky, S., et al.: Inverse design in nanophotonics. Nat. Photonics. **12**(11), 659–670 (2018)
8. Jensen, J.S., Sigmund, O.: Topology optimization for nano-photonics. Laser Photonics Rev. **5**(2), 308–321 (2011)
9. Bendsoe, M.P., Sigmund, O.: Topology Optimization: Theory, Methods, and Applications. Springer, Berlin (2003)
10. Sigmund, O.: On the usefulness of non-gradient approaches in topology optimization. Struct. Multidiscip. Optim. **43**(5), 589–596 (2011)
11. Joannopoulos, J.D., et al.: Photonic Crystals: Molding the Flow of Light, 2nd edn. Princeton University Press, Princeton (2008)
12. Schäferling, M., et al.: Tailoring enhanced optical chirality: design principles for chiral plasmonic nanostructures. Phys. Rev. X. **2**(3), 031010 (2012)
13. Kauranen, M., Zayats, A.V.: Nonlinear plasmonics. Nat. Photonics. **6**(11), 737–748 (2012)
14. Zaabab, A.H., Qi-Jun, Z., Nakhla, M.: A neural network modeling approach to circuit optimization and statistical design. IEEE Trans. Microw. Theory Tech. **43**(6), 1349–1358 (1995)
15. Watson, P.M., Gupta, K.C.: Design and optimization of CPW circuits using EM-ANN models for CPW components. IEEE Trans. Microw. Theory Tech. **45**(12), 2515–2523 (1997)
16. Fang, W., Devabhaktuni, V.K., Qi-Jun, Z.: A hierarchical neural network approach to the development of a library of neural models for microwave design. IEEE Trans. Microw. Theory Tech. **46**(12), 2391–2403 (1998)
17. Burrascano, P., Fiori, S., Mongiardo, M.: A review of artificial neural networks applications in microwave computer-aided design. Int. J. RF Microw. Comput. Aided Eng. **9**(3), 158–174 (1999)
18. Silva, P.H.D.F., Cruz, R.M.S., Assunção, A.G.D.: Blending PSO and ANN for optimal design of FSS filters with Koch Island patch elements. IEEE Trans. Magn. **46**(8), 3010–3013 (2010)
19. Christodoulou, C., Georgiopoulos, M.: Applications of Neural Networks in Electromagnetics. Artech House, Inc, Boston (2000)
20. Kabir, H., et al.: Smart modeling of microwave devices. IEEE Microw. Mag. **11**(3), 105–118 (2010)
21. Andrawis, R.R., et al.: Artificial neural network modeling of plasmonic transmission lines. Appl. Opt. **55**(10), 2780–2790 (2016)

22. Fornarelli, G., et al.: A neural network model of erbium-doped photonic crystal fibre amplifiers. Opt. Laser Technol. **41**(5), 580–585 (2009)
23. Hameed, M.F.O., et al.: Accurate radial basis function based neural network approach for analysis of photonic crystal fibers. Opt. Quant. Electron. **40**(11), 891 (2009)
24. Malheiros-Silveira, G.N., Hernandez-Figueroa, H.E.: Prediction of dispersion relation and PBGs in 2-D PCs by using artificial neural networks. IEEE Photonic. Technol. Lett. **24**(20), 1799–1801 (2012)
25. Wiecha, P.R., Muskens, O.L.: Deep learning meets nanophotonics: a generalized accurate predictor for near fields and far fields of arbitrary 3D nanostructures. Nano Lett. **20**(1), 329–338 (2020)
26. Liu, D., et al.: Training deep neural networks for the inverse design of nanophotonic structures. ACS Photonics. **5**(4), 1365–1369 (2018)
27. Unni, R., Yao, K., Zheng, Y.: Deep convolutional mixture density network for inverse design of layered photonic structures. ACS Photonics. **7**(10), 2703–2712 (2020)
28. Bishop, C.M.: Mixture Density Networks. Aston University: Neural Computing Research Group, Birmingham (1994)
29. Unni, R., et al.: A mixture-density-based tandem optimization network for on-demand inverse design of thin-film high reflectors. Nanophotonics. **10**(16), 4057–4065 (2021)
30. Liu, Z., et al.: Generative model for the inverse design of metasurfaces. Nano Lett. **18**(10), 6570–6576 (2018)
31. Ma, W., et al.: Probabilistic representation and inverse design of metamaterials based on a deep generative model with semi-supervised learning strategy. Adv. Mater. **31**(35), 1901111 (2019)
32. Kudyshev, Z.A., et al.: Machine-learning-assisted metasurface design for high-efficiency thermal emitter optimization. Appl. Phys. Rev. **7**(2), 021407 (2020)
33. Knittl, Z.: Optics of Thin Films: An Optical Multilayer Theory. John Wiley & Sons, Norderstedt (1976)
34. Willey, R.R.: Practical Design and Production of Optical Thin Films, 2nd edn. CRC Press (2002)
35. Macleod, H.A.: Thin-Film Optical Filters, 4th edn. CRC Press, Boca Raton (2010)
36. Kogelnik, H., Shank, C.V.: Stimulated emission in a periodic structure. Appl. Phys. Lett. **18**(4), 152–154 (1971)
37. Shyh, W.: Principles of distributed feedback and distributed Bragg-reflector lasers. IEEE J. Quantum Electron. **10**(4), 413–427 (1974)
38. Born, M., Wolf, E.: Principles of Optics: Electromagnetic Theory of Propagation, Interference and Diffraction of Light, 7th edn. Cambridge University Press, Cambridge (1999)
39. Pilozzi, L., et al.: Machine learning inverse problem for topological photonics. Commun. Phys. **1**(1), 57 (2018)
40. So, S., et al.: Inverse design of ultra-narrowband selective thermal emitters designed by artificial neural networks. Opt. Mater. Express. **11**(7), 1863–1873 (2021)
41. Jiang, J., Fan, J.A.: Multiobjective and categorical global optimization of photonic structures based on ResNet generative neural networks. Nanophotonics. **10**(1), 361–369 (2021)
42. Zhang, D., et al.: Inverse design of an optical film filter by a recurrent neural adjoint method: an example for a solar simulator. J. Opt. Soc. Am. B. **38**(6), 1814–1821 (2021)
43. Wang, H., et al.: Automated multi-layer optical design via deep reinforcement learning. Mach. Learn. Sci. Technol. **2**(2), 025013 (2021)
44. Hegde, R.: Sample-efficient deep learning for accelerating photonic inverse design. OSA Continuum. **4**(3), 1019–1033 (2021)
45. Yin, X., et al.: Terrestrial radiative cooling: using the cold universe as a renewable and sustainable energy source. Science. **370**(6518), 786–791 (2020)
46. Alù, A., Engheta, N.: Multifrequency optical invisibility cloak with layered plasmonic shells. Phys. Rev. Lett. **100**(11), 113901 (2008)
47. Peurifoy, J., et al.: Nanophotonic particle simulation and inverse design using artificial neural networks. Sci. Adv. **4**(6), eaar4206 (2018)
48. So, S., Mun, J., Rho, J.: Simultaneous inverse design of materials and structures via deep learning: demonstration of dipole resonance engineering using core–shell nanoparticles. ACS Appl. Mater. Interfaces. **11**(27), 24264–24268 (2019)

49. Malkiel, I., et al.: Plasmonic nanostructure design and characterization via deep learning. Light Sci. Appl. **7**(1), 60 (2018)
50. Inampudi, S., Mosallaei, H.: Neural network based design of metagratings. Appl. Phys. Lett. **112**(24), 241102 (2018)
51. Nadell, C.C., et al.: Deep learning for accelerated all-dielectric metasurface design. Opt. Express. **27**(20), 27523–27535 (2019)
52. Kiarashinejad, Y., Abdollahramezani, S., Adibi, A.: Deep learning approach based on dimensionality reduction for designing electromagnetic nanostructures. npj Comput. Mater. **6**(1), 12 (2020)
53. An, S., et al.: A deep learning approach for objective-driven all-dielectric metasurface design. ACS Photonics. **6**(12), 3196–3207 (2019)
54. Tanriover, I., Hadibrata, W., Aydin, K.: Physics-based approach for a neural networks enabled design of all-dielectric metasurfaces. ACS Photonics. **7**(8), 1957–1964 (2020)
55. So, S., Rho, J.: Designing nanophotonic structures using conditional deep convolutional generative adversarial networks. Nanophotonics. **8**(7), 1255–1261 (2019)
56. Tang, Y., et al.: Generative deep learning model for inverse design of Integrated nanophotonic devices. Laser Photonics Rev. **14**(12), 2000287 (2020)
57. Yeung, C., et al.: Elucidating the behavior of nanophotonic structures through explainable machine learning algorithms. ACS Photonics. **7**(8), 2309–2318 (2020)
58. Ma, W., Cheng, F., Liu, Y.: Deep-learning-enabled on-demand design of chiral metamaterials. ACS Nano. **12**(6), 6326–6334 (2018)
59. Collins, J.T., et al.: Chirality and chiroptical effects in metal nanostructures: fundamentals and current trends. Adv. Opt. Mater. **5**(16), 1700182 (2017)
60. Plum, E., Fedotov, V.A., Zheludev, N.I.: Extrinsic electromagnetic chirality in metamaterials. J. Opt. A Pure Appl. Opt. **11**(7), 074009 (2009)
61. Ganaie, M.A., et al., *Ensemble deep learning: A review*. Engineering Applications of Artificial Intelligence. **115**, 105151 (2022)
62. Wolpert, D.H.: Stacked generalization. Neural Netw. **5**(2), 241–259 (1992)
63. Liu, Z., et al.: Compounding meta-atoms into metamolecules with hybrid artificial intelligence techniques. Adv. Mater. **32**(6), 1904790 (2020)
64. Georgieva, N.K., et al.: Feasible adjoint sensitivity technique for EM design optimization. IEEE Trans. Microw. Theory Tech. **50**(12), 2751–2758 (2002)
65. Miller, O.D.: Photonic Design: From Fundamental Solar Cell Physics to Computational Inverse Design. University of California, Berkeley (2012)
66. Lalau-Keraly, C.M., et al.: Adjoint shape optimization applied to electromagnetic design. Opt. Express. **21**(18), 21693–21701 (2013)
67. Jiang, J., Fan, J.A.: Global optimization of dielectric metasurfaces using a physics-driven neural network. Nano Lett. **19**(8), 5366–5372 (2019)
68. Jiang, J., et al.: Free-form diffractive metagrating design based on generative adversarial networks. ACS Nano. **13**(8), 8872–8878 (2019)
69. Wen, F., Jiang, J., Fan, J.A.: Robust freeform metasurface design based on progressively growing generative networks. ACS Photonics. **7**(8), 2098–2104 (2020)
70. Makhzani, A. et al., Adversarial autoencoders. arXiv preprint arXiv:1511.05644 (2015)
71. Ra'di, Y., Simovski, C.R., Tretyakov, S.A.: Thin perfect absorbers for electromagnetic waves: theory, design, and realizations. Phys. Rev. Appl. **3**(3), 037001 (2015)
72. Watts, C.M., Liu, X., Padilla, W.J.: Metamaterial electromagnetic wave absorbers. Adv. Mater. **24**(23), OP98–OP120 (2012)
73. Cui, Y., et al.: Plasmonic and metamaterial structures as electromagnetic absorbers. Laser Photonics Rev. **8**(4), 495–520 (2014)
74. Naik, G.V., Shalaev, V.M., Boltasseva, A.: Alternative plasmonic materials: beyond gold and silver. Adv. Mater. **25**(24), 3264–3294 (2013)
75. Woldseth, R.V., et al., *On the use of artificial neural networks in topology optimisation*. Structural and Multidisciplinary Optimization. **65**(10), 294 (2022)

Chapter 5
Deep-Learning-Enabled Applications in Nanophotonics

Abstract The data-driven nature of DL makes it well suited for exploring the enormous design space in a different way than conventional optimization techniques do. Nonetheless, the possible usage of DL in optics and photonics can be much broader, going far beyond inverse design tasks. This chapter will focus on these applications. Specifically, we will discuss two types of problems: prediction of optical properties and interpretation of optical data. Examples are selected to feature representative DL methods and nanophotonic applications. The first two sections are dedicated to surrogate simulators, which, respectively, showcase the utility of transfer learning and the capability of rapid modeling of complex near fields in a variety of nanostructures. Section 5.3 discusses the use of nanostructures for information encoding and its readout by an ANN.

5.1 Knowledge Migration by Transfer Learning

We start the discussion with a surrogate simulator used in scenarios similar to those in Sect. 4.2. As introduced earlier, the function of a surrogate simulator is to predict the optical responses for a class of nanostructures [1]. Having no concern of non-uniqueness in forward problems, the choice of DL model for building a surrogate simulator is as naive as a feedforward fully connected NN, and other than a handful of hyperparameters subject to tuning, there seems to have no reason for further discussion. Whereas it is true that training a surrogate simulator is a relatively simple task most of the time, the prerequisite condition of a sufficient data supply should not be forgotten. As seen in the last chapter, typically at least tens of thousands of samples are required for training to reach desirable performance. In many applications, generating or collecting such a large amount of data is not quite feasible or even possible. Taking inverse design, for example, the analytical solution of planar multilayers ensures that massive data generation is computationally cheap, but this is not the case for metamaterials and metasurfaces [2–6]. Moreover, a NN can only handle a very specific type of nanostructures. For instance, in Sect. 4.2, the number of layers and choice of materials are both fixed in each study. If one wonders if an

K. Yao, Y. Zheng, *Nanophotonics and Machine Learning*, Springer Series in Optical Sciences 241, https://doi.org/10.1007/978-3-031-20473-9_5

18- or 19-layer stack can totally resemble a 20-layer high reflector, or if a 21- or 22-layer stack can further extend the high-reflection zone, the existing model trained for 20-layer thin films will have troubles in addressing these queries directly. A lazy yet somewhat painful solution is to train a different model for each number of layers at the cost of recreating new datasets, unless this factor is included in the design parameters.

On the other hand, from human intuition, adding or removing a few layers from a succession of thin films does alter the achievable optical responses but does not change the physical principles involved, which can be clearly seen from the formalism of transfer matrix method [7–9]. Light is partly reflected at each interface of the multilayers in a fashion determined by the two materials forming that interface, and the total effect is just a consequence of interference. In other words, the problems of solving the optical properties of multilayer structures that differ in layer numbers are related. Some knowledge corresponding to the common underlying physical principles, although manifested in forms different from the analytical expressions, can likely be shared in DL implementations as well. And if applicable, this will significantly improve the sample efficiency of DL algorithms.

Transfer learning could provide a practical solution in this situation [11, 12]. Figure 5.1a illustrates the generic process, which takes place between two DL models. The first model aims at solving a problem for which a large dataset is available. Its training can be accomplished by following the standard workflow. The other model for a related task, by contrast, has very limited training data due to great expense of sample creation or collection, suffering a high risk of poor performance or overfitting. Transfer learning attempts to apply the knowledge gained during the training of the first model (source model) to the other one (target model), thereby improving the training of the latter. Going back to the example of multilayer thin

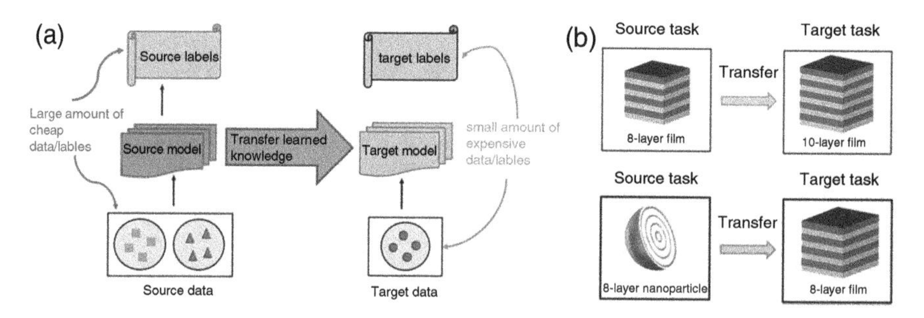

Fig. 5.1 (**a**) Transfer learning between a source domain (left) and a target domain (right). Knowledge learned from a source model is used to ease or improve the training of a target model. The latter may suffer from a small dataset, leading to poor performance. (**b**) Examples of forward optical problems which transfer learning can be applied to. Upper row: Source and target tasks are similar, e.g., solving the transmittance of planar multilayer stacks with different numbers of layers. Lower row: Source and target tasks are of considerable difference, e.g., solving the scattering cross section of multilayer core-shell particles and the transmittance of planar stacks with the same number of layers. (Reprinted from [10] with permission. Copyright (2019) American Chemical Society)

films, a well-trained surrogate simulator for a certain layer number can be used as a source to assist the training of simulators for structures with different layers, as shown in Fig. 5.1b. In a more general sense, knowledge transfer is possible between different physical problems, which arise from drastically different structures but exhibit optical behaviors governed by some common physical rules, such as scattering by multilayer core-shell particles and transmission/reflection of multilayer thin films [10].

The implementation of transfer learning is usually achieved by reusing some components of the source model in the target model. The portion of the reused parts in the entire model needs to be tuned, and the optimal strategy varies across different situations. Figure 5.2 illustrates the architectures of two surrogate simulators for alternating TiO_2 and SiO_2 thin-film stacks with different numbers of films. Without loss of generality, the two NNs have identical architectures. The input layers taking vectors of film thicknesses have the dimension equal to the larger number of layers between the two stacks in the source and target domains; unused positions are set to 0. The output layers take transmittance values uniformly discretized over the same

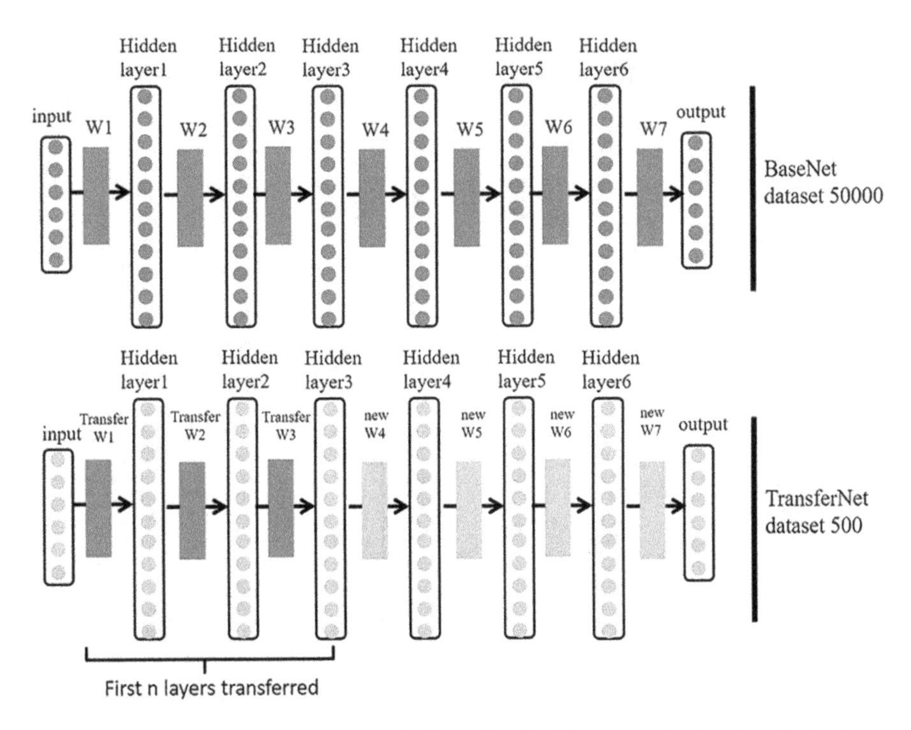

Fig. 5.2 Architectures of two surrogate simulators involved in a case of optical knowledge transfer. Upper: The source model has six hidden layers and is trained from scratch on a large dataset containing 50,000 samples. Lower: The target model dealing with a different forward problem has the same architecture but is trained (fine-tuned) on a small dataset of 500 samples, while the initial weights and biases of the first n hidden layers are duplicated from the source model. (Reprinted from [10] with permission. Copyright (2019) American Chemical Society)

wavelength range and naturally fit the architecture. Six hidden layers are used between the input and output, leading to seven layers (W1–W7) of weights and biases. The source model is trained from scratch over a dataset of 50,000 samples, whereas for the target model, the first few layers of weights and biases are initialized with values copied from the source model and the remaining layers with random values from a normal distribution. To demonstrate the capability of knowledge migration, the target model is intended to be exposed to a truly small dataset of 500 samples. The training is thus a process of fine-tuning of the network parameters.

Figure 5.3a, b shows the performance of this scheme when the source and target models are designed for ten- and eight-layer stacks, respectively, and vice versa. Spectrum error is defined as the spectrally averaged difference between the predicted transmittance and ground truth:

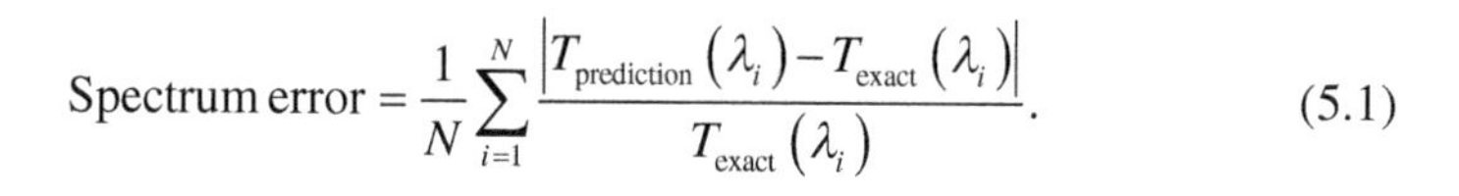

$$\text{Spectrum error} = \frac{1}{N}\sum_{i=1}^{N}\frac{\left|T_{\text{prediction}}(\lambda_i) - T_{\text{exact}}(\lambda_i)\right|}{T_{\text{exact}}(\lambda_i)}. \tag{5.1}$$

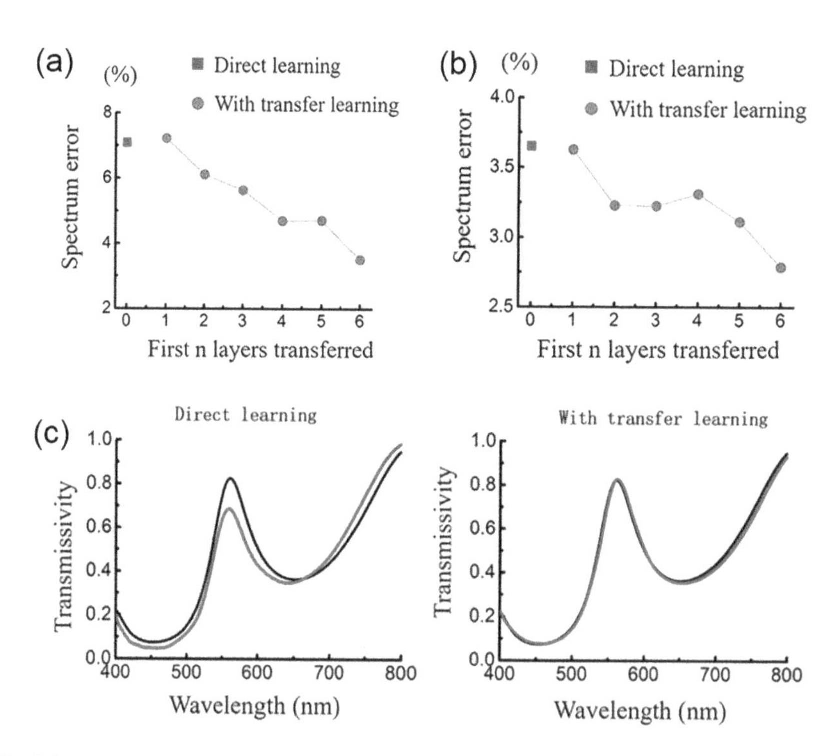

Fig. 5.3 (**a**) Test errors of a target surrogate simulator for computing the transmittance of eight-layer stacks of alternating TiO_2 and SiO_2 thin films, with the first *n* hidden layers transferred from a source surrogate simulator for ten-layer stacks. (**b**) Same as (**a**) but with the source and target switched. (**c**) Transmittance spectra (red curves) predicted by a model trained from scratch on 500 samples (i.e., direct learning) and by a model with transfer learning, in comparison to the ground truth spectrum (black curves) for an eight-layer stack. (Reprinted with permission from [10]. Copyright (2019) American Chemical Society)

Direct learning refers to the case where the target model receives no information from the source and is trained from scratch independently on the 500 samples of its own. For knowledge migration in both directions, noticeable error reduction can be recognized when at least two layers of parameters are duplicated. The improvement in performance is further visualized in Fig. 5.3c with a selected example for the case where knowledge is transferred from ten-layer stacks to eight-layer stacks.

One of the reasons that direct learning based on 500 samples can still make predictions not far away from the ground truth is probably the limited complexity of the forward problem in study. Because the film thicknesses are restricted in the range between 30 and 70 nm, the achievable transmittance spectra may not be diverse enough in line shape. It is reasonable to assume that when the number of layers increases and the thickness ranges are extended, the improvement by transfer learning is more significant. Moreover, duplication of weights does not necessarily start from the first layer. In another trial of knowledge transfer between multilayer core-shell particles and multilayer thin films [10, 13], it is discovered that error reduction is maximal when certain layers in the middle of the NN are copied for initialization, and inclusion of additional layers is either not helpful or even harmful to the final performance.

An interesting extension of transfer learning is the multitask learning. When none of a few related tasks have adequate data or sample efficiencies are critical for these tasks as a whole, one possible solution is to build a model as shown in Fig. 5.4.

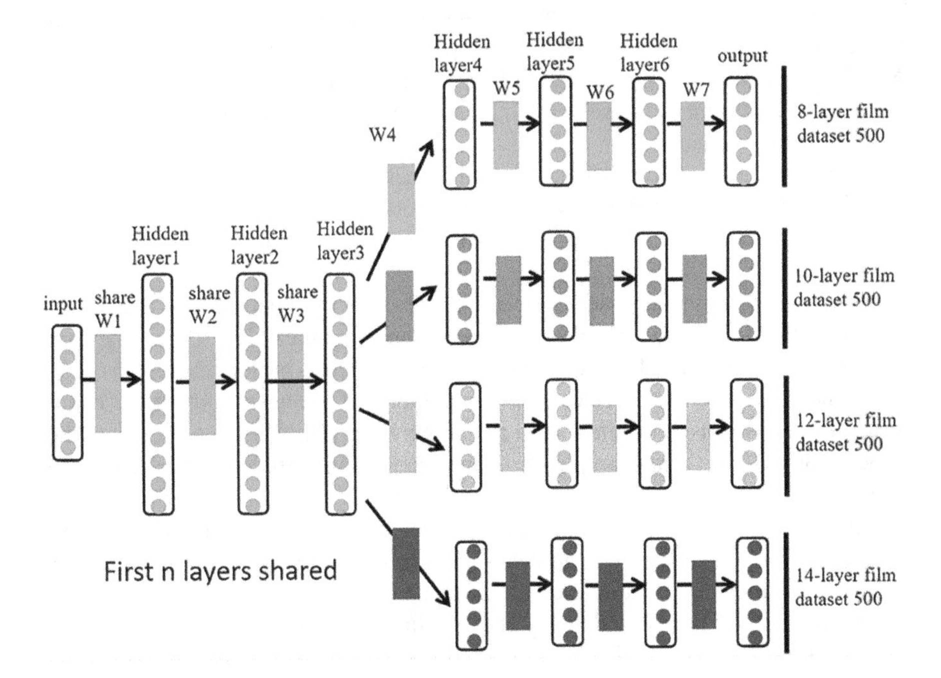

Fig. 5.4 Surrogate simulators based on a common form of multitask learning. (Reprinted from [10] with permission. Copyright (2019) American Chemical Society)

In multitask learning, the tasks share the same input but end up with task-specific outputs. Between these two layers, a few hidden layers containing generic parameters are also shared across all the tasks, while higher layers are split into branches associated with each target domain. The training process needs to be adjusted accordingly, where data are fed in the model as a pool of samples, and during back-propagation, parameters are updated in the shared layers and in task-specific layers which the samples belong to.

Finally, the applications presented in this section usually do not suffer from difficulties in data generation. Multilayer structures are chosen as examples here mainly because discussions can be focused on explaining the DL methods. From a practical point of view, the advantages of transfer learning have not been convincingly proved for complex nanophotonic systems with the pressing need for high sample efficiencies. Some concerned efforts were recently made in this direction for metasurface design [14–16].

5.2 Surrogate Simulators for Inferring Field Quantities

Until now, the surrogate simulators we have discussed, working alone or being a part of an inverse design framework, output optical responses of given structures in the form of a specific observable, such as transmittance or reflectance spectra, scattering patterns, or near-field intensity maps. In nanophotonics, such modeling is usually done by performing full-wave simulations, in which the complete set of Maxwell's equations is numerically solved through spatial and temporal discretization. A real electromagnetic solver computes the vector electric and magnetic fields everywhere in the simulation domain, and aforementioned observables can all be derived from the 3D vector field distributions via some well-defined evaluations and transformations. If a surrogate simulator can totally resemble an electromagnetic solver, it will be a big advance in fast simulation and in generalization of data availability.

Wiecha and coworkers demonstrated a DL-based framework as a general-purpose predictor of the optical responses of nanostructures [17]. The framework comprises two parts. The first one is a homebuilt simulation package based on CDA, which provides a scheme for accurate computation of the near fields of 3D nanostructures as well as convenient data reconfiguration [18]. Figure 5.5 depicts the concept of CDA with an example of a nanosphere. The particle is discretized into small cubic cells centered at $\mathbf{r}_i$ with a dipolar polarizability $\alpha(\mathbf{r}_i, \omega)$. Each cell is assigned an electric dipole moment $\mathbf{p}_i(\omega)$ induced by the local electric field $\mathbf{E}(\mathbf{r}_i, \omega)$, determined self-consistently by all the other induced dipoles $\mathbf{p}_j(\omega)$ and the incident field. The volume discretization can be extended to include the surrounding medium, but no dipoles will be induced there. The coupling between discretized dipoles can be evaluated by using the Green dyadic method, and the final result is the electric polarization distribution inside the particle. In many works, this method is also termed discrete dipole approximation (DDA) [19].

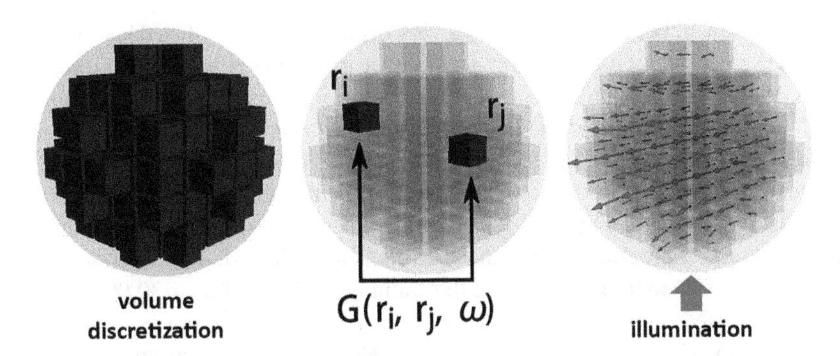

Fig. 5.5 Schematic of coupled dipole approximation in the case of a sphere. Left: The volume of the sphere is discretized into small cubic cells. Central: Each cell is treated as an electric dipole with an effective polarizability. Self-consistent fields are obtained through calculating the coupling between all pairs of cells described by the Green's dyad **G**. Right: Self-consistent electric polarization distribution in the sphere under external illumination. Each small red arrow represents the electric polarization of a cell. (Adapted from [17] with permission. Copyright (2019) American Chemical Society)

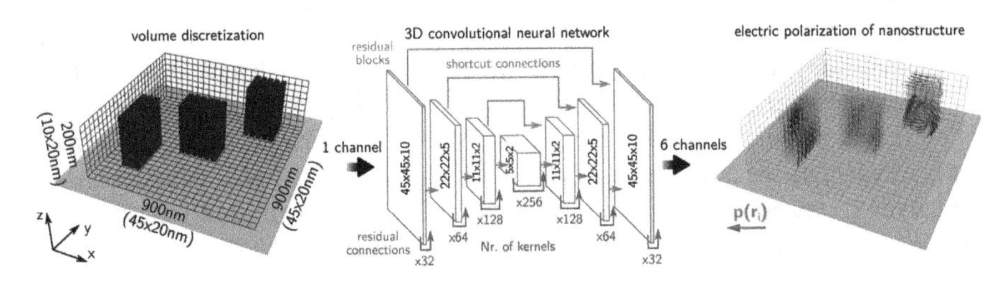

Fig. 5.6 A 3D CNN for predicting the electric fields inside three Si nanocubes on a glass substrate. Digits in the left panel denote the dimensions of the simulation domain and volume discretization scheme (number of cells (mesh grids) × mesh size). Red arrows in the right panel represent the real part of electric field vectors, taken from three of the six output channels. (Reprinted from [17] with permission. Copyright (2019) American Chemical Society)

The second part of the framework is a 3D symmetric fully convolutional NN. As shown in Fig. 5.6, the network receives information about the geometry of the nanostructure and outputs in six channels the complex electric fields (real and imaginary parts of E_x, E_y, and E_z). A nanostructure made of one material is represented by an array of binary digits indicating whether the discrete cells are occupied by the material (1) or not (0), and the predicted fields are given on the same grid of positions. The volume discretization strategy of CDA automatically saves the efforts to reformat data. Other mesh grids that are commonly used in simulation tools, such as tetrahedrons, may require extra conversion between mesh labels and spatial coordinates. The construction of the NN also features two interesting designs. In order to reinforce the ability to reconstruct fine spatial information—in the present case the internal fields—the NN uses an encoder-decoder architecture built with "U-Net"-type shortcut connections between the corresponding down- and up-sampling blocks. Moreover, because the NN has a very deep architecture of ~90 layers, which could cause degradation of training accuracy, it is organized in sequential residual blocks. Within one of such units, the first convolutional layer is added via a shortcut connection to the outputs of the last normalization layer [20]. The idea of residual learning is illustrated in Fig. 5.7 with a standard NN architecture. Intuitively, a residual network offers a solution that improves the model's capability to fit simple mappings (e.g., identity) by adding shortcut connections, while the fitting of complex mappings by the stack of nonlinear layers is not affected.

Despite the 3D nature of the task, the training of this NN uses 30,000 samples each for gold and Si nanostructures, around the same size as in the previous examples. On an NVIDIA P6000 GPU, the training takes about 2 or 10 minutes per epoch for gold and Si datasets, respectively, and it is terminated after 100 epochs as the validation accuracy saturates. Once trained, the 3D surrogate simulator can give predictions in less than 10 ms for one structure. This time increases to some tens or up to hundreds of ms if the hardware is changed to an Intel i7-3770 CPU, which is nonetheless still several orders of magnitude shorter than needed by an electromagnetic solver. It is more interesting to take a tour to the results for Si nanostructures, which can be related to some earlier discussions in Sect. 1.2.

Figure 5.8a summarizes the magnetic responses of Si nanorods under normal incidence of TE and TM polarized plane waves at 700-nm wavelength, predicted by the 3D surrogate simulator (left column) and computed by the CDA package (right column). The magnetic response is characterized by the contribution to the extinction coefficient by the magnetic dipole moment of the nanorod, which does not have an analytical expression as for circular cylinders but can be extracted from the predicted/simulated near fields through multipole expansion [21, 22]. In Fig. 5.8a, decent agreement except small discrepancies of intensity at some points is obtained for both excitations, confirming the accuracy of the NN. In addition to derived physical quantities, a more direct evidence, the electric field distributions inside a resonant nanorod are also presented. The circulating vectors perpendicular to the rod's major axis evidently reveal the profile of a magnetic dipolar mode under TE-polarized

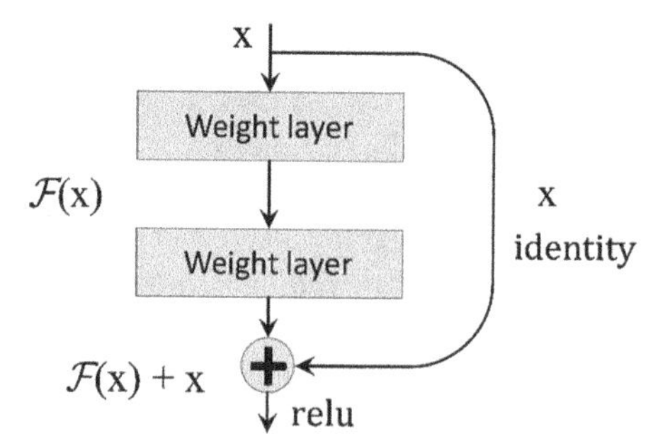

Fig. 5.7 Sketch of a residual block. The input x is added to the output of a deeper layer via a shortcut connection, recasting the original desired mapping $H(x)$ to the form $F(x) + x$. For a very deep architecture, optimization of the residual mapping $F(x)$ is proved easier than of the original, unreferenced mapping $H(x)$

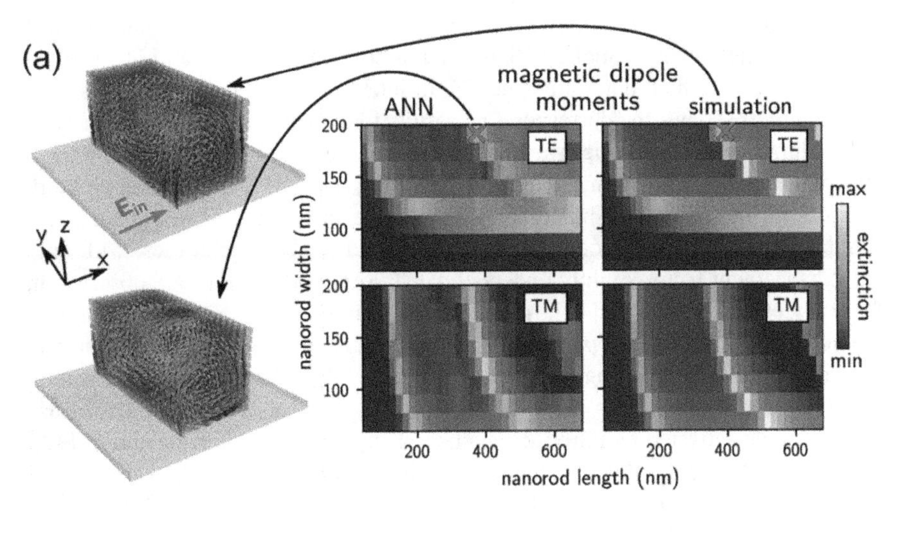

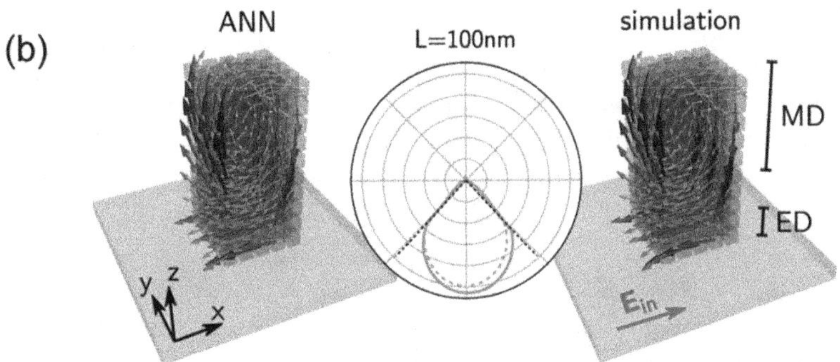

Fig. 5.8 Reproducibility of complex near fields in Si nanorods and derived optical effects in the far field. (**a**) Maps of contributions to the extinction coefficient by the magnetic dipole moment of a nanorod, as functions of its length and width for TE and TM polarized excitations. (**b**) Electric fields inside a cuboid (100 × 100 × 200 nm^3) and the scattering patterns by ANN prediction (solid line) and by simulation (dashed line). The height of the nanostructures is 200 nm, and the wavelength is fixed at 700 nm. (Reprinted from [17] with permission. Copyright (2019) American Chemical Society)

illumination. Similar analysis can be performed for electric responses. On top of this, the geometric condition for directional scattering, a consequence of the Kerker condition [23], is found for the given thickness of nanostructures and wavelength. The predicted and simulated field distributions and scattering patterns are shown in Fig. 5.8b. While we have seen the far-field patterns in Sects. 2.1.2 and 4.2.2, the field distributions visualize some important aspects of the Kerker condition on dipole orientations and strengths, which states that ED and MD must be perpendicular to each other and have comparable magnitudes.

Accurate prediction of the near fields not only allows for easy derivation of multiple far-field behaviors but also enables the evaluation of mutual interactions between coupled nanostructures. Two representative quantities of particular interest are the near-field enhancement, which has a direct relation to LDOS [24], and the optical chirality, which determines how light interacts with chiral substances [25–27]. Figure 5.9 shows the corresponding results for Si nanocube dimers. In a symmetric dimer, when the incident field is polarized along the dimer axis, a well-known effect is the generation of an electric hotspot in the gap [28], while optical chirality vanishes on the symmetry planes or after volume integration. The 3D surrogate simulator successfully reproduces the correct trend for both quantities, with only the enhancement factor slightly underestimated. Similar deviations are observed in the study of gold nanorods [17]. In fact, for DL-based simulators, it is common to see noticeable errors near optical resonances, because they are relatively sparse in the dataset. Besides changing the excitation to CPL [29] or twisting the polarization direction off the dimer axis [29–31], the vanishing optical chirality in symmetric dimers can be activated by introducing a lateral shift between the two cuboids [32]. As shown in Fig. 5.9b, the dependence of chirality magnitude on the gap size and shift and the correlation between handedness and the direction of the shift are predicted with reasonable accuracy.

The implementation of 3D surrogate simulators requires sophisticated and likely task-specific models, plus extra configuration of data format. Efforts along this line are currently rare, probably because of the knowledge and skill gaps for researchers in the nanophotonic community. It is not clear yet if and how much an affordable increase in data size will improve the performance, especially for nanostructures

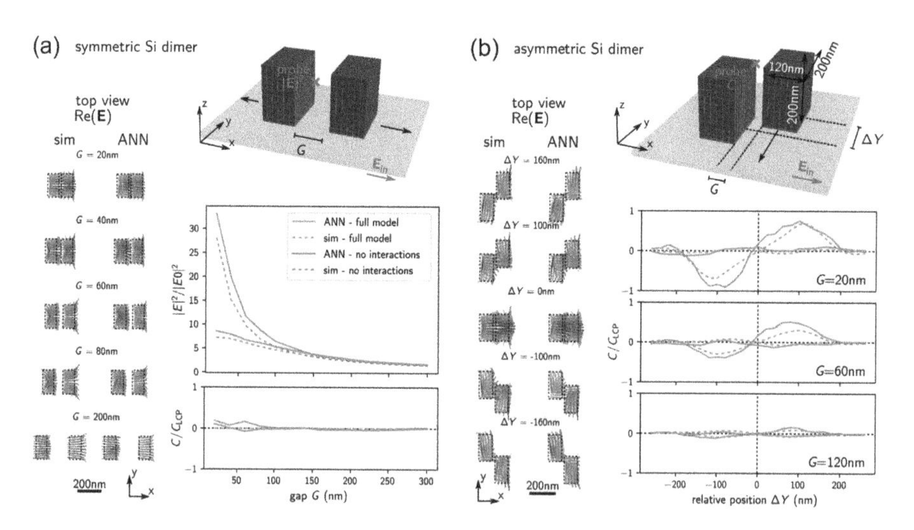

Fig. 5.9 Reproducibility of near-field enhancement and optical chirality for Si nanocube dimers formed by two identical cuboids ($120 \times 200 \times 200$ nm^3). (**a**) Symmetric dimers with different gap sizes. (**b**) Asymmetric dimers with varying gap sizes and lateral offsets. For all structures, excitations are x-polarized plane waves normally incident from the top. Orange curves are obtained by superposing the fields from two separate models, each containing one of the two cuboids in the original location. (Reprinted from [17] with permission. Copyright (2019) American Chemical Society)

under resonance conditions. The CDA simulation of in total 60,000 samples used in [17] took about 10 days on two workstations. Strategies and techniques for speed-up or smarter sampling will be helpful for further study of this topic.

5.3 Interpretation of Optical Information

Solving inverse problems is a process of seeking the underlying mapping from optical responses to physical variables. In inverse design tasks, the latter are the design parameters, and the one-to-many problem must be suitably taken care of, as discussed in Sect. 4.1. But supposing there is a subset of inverse problems, where the optical responses and physical structures have a one-to-one correspondence, training a DL model to retrieve the structures would be as easy as training a forward surrogate solver. Moreover, viewing from the standpoint of applications, uniquely determining a structure by some of its properties is just what information decoding does, and if the property is an optical response, it is the optical information stored in the structure being decoded.

A very interesting demonstration of high-density optical data storage and associated readout scheme is reported by Wiecha and coworkers [33]. The idea combines the advantages from both sides of nanophotonics and DL to address the challenges in information storage. Let us start by checking the feasibility of encoding information into nanostructures. In traditional optical storage techniques like compact disks, data bits are stored on some media in the form of pits [34]. The smallest dimensions of individual pits and the areas between them, termed lands, are diffraction-limited to ensure that two distinct reflection states can be reliably achieved. Increasing the density of pits suffers from fundamental difficulties in writing and optical readout. Nanophotonics provides a viable solution to a part of these challenges [35]. Subwavelength nanostructures exhibit rich scattering behaviors. In the first two chapters, we have seen many examples. The next question is "how can binary data be encoded into a nanostructure?". Figure 5.10 illustrates one possible scheme, which is designated for representing 9-bit information. Instead of using discrete values of the geometric parameters, here the structures are treated as patterns. Nine deep-subwavelength cuboids are organized into 3 × 3 grids, each occupied by Si (digit "1") or air (digit "0"). To make symmetric geometries distinguishable in polarized spectroscopy, an L-shaped sidewall is added, while the whole structure is still subwavelength in all dimensions. Nine-bit structures have $2^9 = 512$ possible variations. Ideally, each of them needs to possess a unique scattering spectrum to comprise an encoding system. An effective approach to quantifying and visualizing the uniqueness of spectra is the t-SNE [36], which displays similar data by nearby points and dissimilar ones by separated points. Figure 5.11 shows the layouts of nanostructures for encoding sequences of 2–5 bits and the corresponding t-SNE plots for experimentally collected spectra of hundreds of fabricated samples. Three types of spectra are analyzed: x-polarized scattering, y-polarized scattering, and using both as one set, denoted by XY polarization. These datasets behave similarly

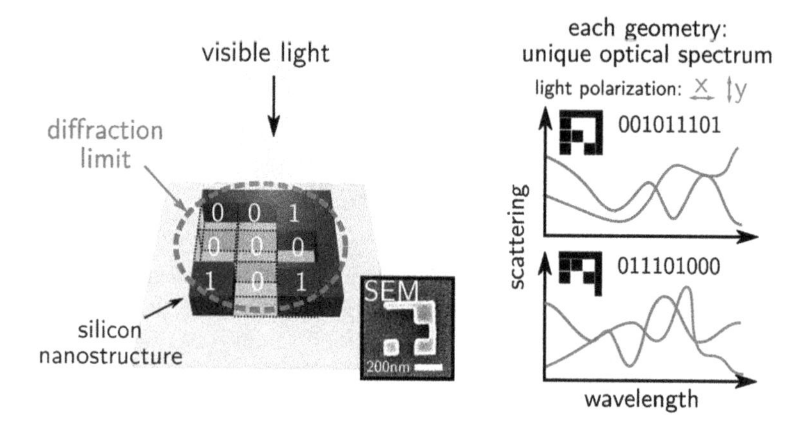

Fig. 5.10 Encoding information in nanostructures. Left: 9-bit information is encoded in a Si nanostructure consisting of nine cubic blocks. An L-shaped sidewall is introduced to distinguish symmetric geometries in polarized spectroscopy. The entire structure is subwavelength in size. Right: Examples to show the one-to-one correspondence between each geometry and a unique scattering spectrum. Curves in blue and orange are from x- and y-polarized light, respectively. (Reprinted from [33] with permission of Springer Nature)

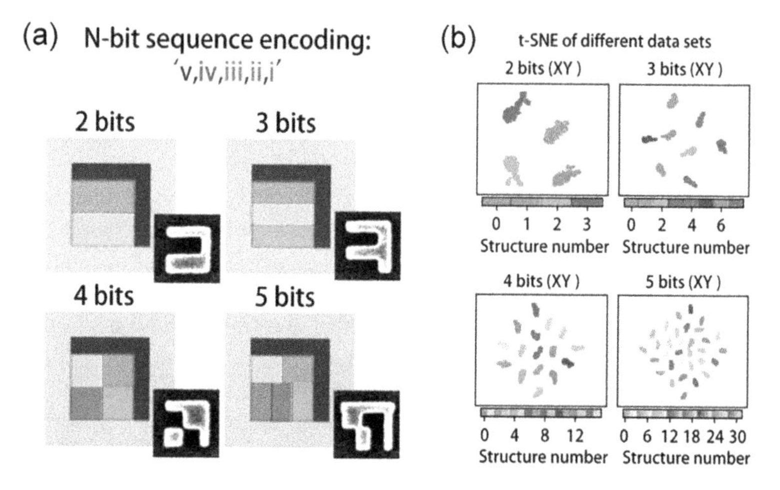

Fig. 5.11 (**a**) Schemes for encoding 2–5 binary digits in Si nanostructures. Bit sequences are indicated by color coding. Inset SEM images show areas of 450 × 450 nm^2 containing selected samples for "10," "010," "1001," and "01010," respectively. (**b**) t-SNE plots of the *XY* polarization training datasets for structures in (a), showing clear clustering of spectra and in turn identification of structures. (Reprinted from [33] with permission of Springer Nature)

for cases of 2–4 bits, while for longer strings, only the *XY* polarization sets can still be well separated into clusters, as shown in Fig. 5.11b.

Once confirming the one-to-one correspondence between the nanostructures and their spectra, the next thing to resolve is decoding the stored information, i.e., retrieval of the bit sequence from a measured spectrum. This task is much easier for

DL than inverse design. The architecture of the NN is chosen to be a 1D CNN, followed by a fully connected network (Fig. 5.12a). Compared to the surrogate simulators of spectra that only have the latter part, the introduction of a CNN enhances the model's ability to conduct pattern recognition. The training and subsequent readout operation of the network are straightforward. Possible tricks, if any, are in the data collection phase. Because the training data are experimentally measured backscattering spectra, the total amount of work needed for larger numbers of digits, e.g., 9-bit sequences, is too massive for a proof-of-concept demonstration. As a compromise, an artificially extended dataset is generated based on measurements from 4 copies of the 512 different geometries. Briefly, for each geometry, semi-experimental data $\sigma(\lambda)$ is created by multiplying the weighted sum of the four measured spectra $\sigma_i(\lambda)$ by a random scaling factor C ($0.9 < C < 1.1$):

$$\sigma(\lambda) = C\sum_{i=1}^{4} w_i \sigma_i(\lambda), \tag{5.2}$$

under the constraint that the random weights w_i add up to unity. Like symmetry is utilized in the extension of datasets for inverse design of metasurfaces [6], the reasoning of Eq. (5.2) can be based on the fact that variations between spectra of different copies mainly arise from fabrication errors and light intensity fluctuations. Using this method, the dataset is expanded to having 300 samples per geometry,

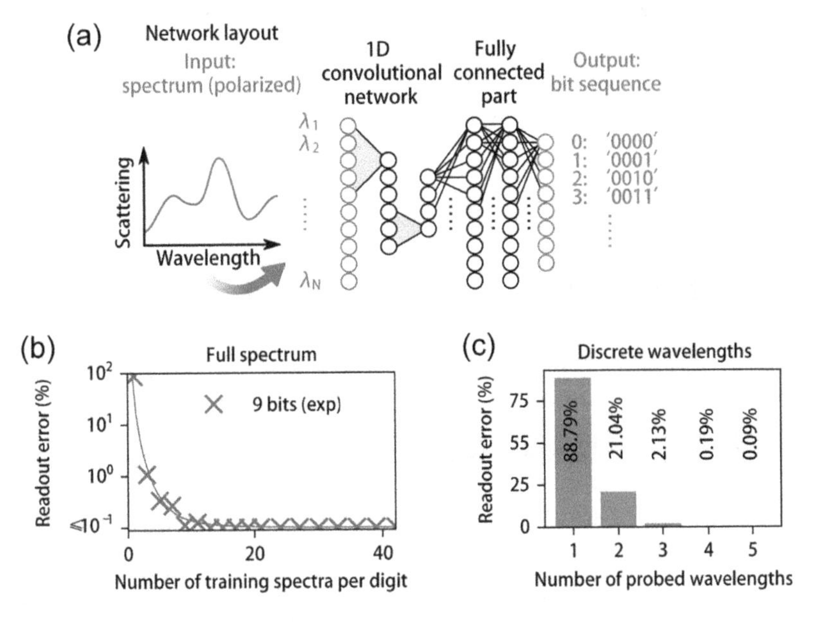

Fig. 5.12 (**a**) Architecture of the NN used for spectra classification. (**b**, **c**) Accuracy of NNs trained on an artificially extended 9-bit experimental dataset using full spectra of backscattering (**b**) or backscattering intensities at discrete wavelengths between 500 and 740 nm (**c**). (Reprinted from [33] with permission of Springer Nature)

sufficient for representing the underlying structure. Figure 5.12b shows the readout accuracy of a trained NN. Almost error-free performance is achieved.

Using full spectra for readout is practically not convenient. It is of great interest to relax the requirements on optical measurements as much as possible. A simple improvement is to instead use scattering intensities at a handful of discrete wavelengths. This simplification turns out to be sensitive to the number of probed wavelengths. As compared in Fig. 5.12c, sampling the spectra at one or two wavelengths leads to very large readout error, while the accuracy is recovered quickly as additional data points are added. An implicit limitation of this method is that spectra cannot be sampled at arbitrary wavelengths. Some prior knowledge about the spectra is needed to ensure that the probed intensities of different samples are distinguishable, which can be analyzed by using t-SNE.

An even more aggressive simplification of the readout is made to get rid of any spectroscopy. The hypothesis is that spectral information after integration over fixed wavelength intervals still contains enough representations of the scattering properties, which can otherwise be acquired via imaging. Figure 5.13 shows the readout scheme based on RGB information of the sample images. Compared to the previous methods using spectroscopic data, the new scheme needs only several RGB values. This change not only simplifies the network structure, but, more importantly, provides a possible solution to the massively parallel readout of the encoded data [33]. In addition to the RGB values of polarization-filtered images, the scattering intensity that contains independent information can be added to the input vector. This strategy proves effective in improving accuracy. For the 4-bit sequences, the readout is quasi-error-free.

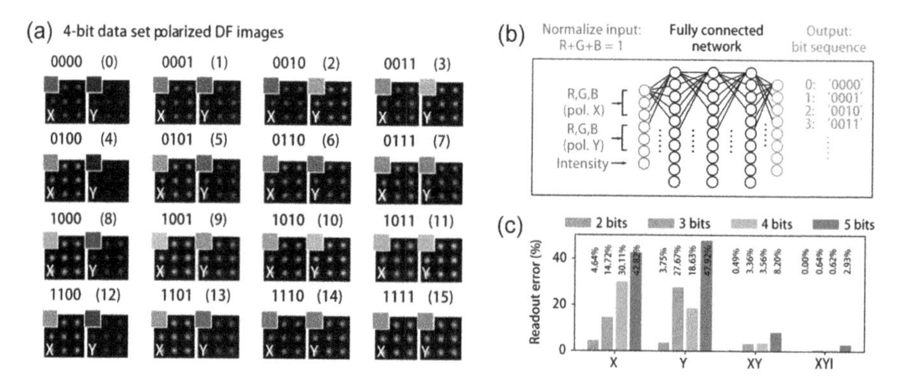

Fig. 5.13 (**a**) Library of polarization-filtered dark-field images of 3 × 3 arrays of 4-bit nanostructures. In each image, the inset box on the upper left has an RGB color averaged over the 9 units, and the letter on the lower left denotes polarization. (**b**) Architecture of the NN for RGB color classification. (**c**) Readout accuracy of NNs for cases of 2–5 bits, trained on different datasets consisting of RGB and intensity values. *X*, *X*-filtered; *Y*, *Y*-filtered; *XY*, *X*-filtered and *Y*-filtered; *XYI*, *XY* plus scattering intensity *I*. (Reprinted from [33] with permission of Springer Nature)

References

1. Jiang, J., Chen, M., Fan, J.A.: Deep neural networks for the evaluation and design of photonic devices. Nat. Rev. Mater. **6**(8), 679–700 (2021)
2. Liu, D., et al.: Training deep neural networks for the inverse design of nanophotonic structures. ACS Photonics. **5**(4), 1365–1369 (2018)
3. Unni, R., et al.: A mixture-density-based tandem optimization network for on-demand inverse design of thin-film high reflectors. Nanophotonics. **10**(16), 4057–4065 (2021)
4. Jiang, J., Fan, J.A.: Multiobjective and categorical global optimization of photonic structures based on ResNet generative neural networks. Nanophotonics. **10**(1), 361–369 (2021)
5. Malkiel, I., et al.: Plasmonic nanostructure design and characterization via deep learning. Light Sci. Appl. **7**(1), 60 (2018)
6. Kudyshev, Z.A., et al.: Machine-learning-assisted metasurface design for high-efficiency thermal emitter optimization. Appl. Phys. Rev. **7**(2), 021407 (2020)
7. Born, M., Wolf, E.: Principles of Optics: Electromagnetic Theory of Propagation, Interference and Diffraction of Light, 7th edn. Cambridge University Press, Cambridge (1999)
8. Knittl, Z.: Optics of Thin Films: An Optical Multilayer Theory. John Wiley & Sons (1976)
9. Katsidis, C.C., Siapkas, D.I.: General transfer-matrix method for optical multilayer systems with coherent, partially coherent, and incoherent interference. Appl. Opt. **41**(19), 3978–3987 (2002)
10. Qu, Y., et al.: Migrating knowledge between physical scenarios based on artificial neural networks. ACS Photonics. **6**(5), 1168–1174 (2019)
11. Yosinski, J., et al.: How transferable are features in deep neural networks? In: Advances in Neural Information Processing Systems 27 (NIPS 2014)
12. Weiss, K., Khoshgoftaar, T.M., Wang, D.: A survey of transfer learning. J. Big Data. **3**(1), 9 (2016)
13. Qiu, C., et al.: Nanophotonic inverse design with deep neural networks based on knowledge transfer using imbalanced datasets. Opt. Express. **29**(18), 28406–28415 (2021)
14. Xu, D., et al.: Efficient design of a dielectric metasurface with transfer learning and genetic algorithm. Opt. Mater. Express. **11**(7), 1852–1862 (2021)
15. Zhu, R., et al.: Phase-to-pattern inverse design paradigm for fast realization of functional metasurfaces via transfer learning. Nat. Commun. **12**(1), 2974 (2021)
16. Zhang, J., et al.: Heterogeneous transfer-learning-enabled diverse metasurface design. Adv. Opt. Mater. **10**(17), 2200748 (2022)
17. Wiecha, P.R., Muskens, O.L.: Deep learning meets nanophotonics: a generalized accurate predictor for near fields and far fields of arbitrary 3D nanostructures. Nano Lett. **20**(1), 329–338 (2020)
18. Girard, C.: Near fields in nanostructures. Rep. Prog. Phys. **68**(8), 1883–1933 (2005)
19. Draine, B.T., Flatau, P.J.: Discrete-dipole approximation for scattering calculations. J. Opt. Soc. Am. A. **11**(4), 1491–1499 (1994)
20. He, K., et al.: Deep residual learning for image recognition. In: 2016 IEEE Conference on Computer Vision and Pattern Recognition (CVPR) (2016)
21. Jackson, J.D.: Classical Electrodynamics, 3rd edn. John Wiley & Sons, New York (1999)
22. Alaee, R., Rockstuhl, C., Fernandez-Corbaton, I.: An electromagnetic multipole expansion beyond the long-wavelength approximation. Opt. Commun. **407**, 17–21 (2018)
23. Liu, W., Kivshar, Y.S.: Generalized Kerker effects in nanophotonics and meta-optics. Opt. Express. **26**(10), 13085–13105 (2018)
24. Novotny, L., Hecht, B.: Principles of Nano-optics, 2nd edn. Cambridge University Press, Cambridge (2012)
25. Lipkin, D.M.: Existence of a new conservation law in electromagnetic theory. J. Math. Phys. **5**(5), 696–700 (1964)
26. Tang, Y., Cohen, A.E.: Optical chirality and its interaction with matter. Phys. Rev. Lett. **104**(16), 163901 (2010)

27. Tang, Y., Cohen, A.E.: Enhanced enantioselectivity in excitation of chiral molecules by superchiral light. Science. **332**(6027), 333–336 (2011)
28. Albella, P., et al.: Low-loss electric and magnetic field-enhanced spectroscopy with subwavelength silicon dimers. J. Phys. Chem. C. **117**(26), 13573–13584 (2013)
29. Yao, K., Liu, Y.: Enhancing circular dichroism by chiral hotspots in silicon nanocube dimers. Nanoscale. **10**(18), 8779–8786 (2018)
30. Zhao, X., Reinhard, B.M.: Switchable chiroptical hot-spots in silicon nanodisk dimers. ACS Photonics. **6**(8), 1981–1989 (2019)
31. Tian, X., Fang, Y., Sun, M.: Formation of enhanced uniform chiral fields in symmetric dimer nanostructures. Sci. Rep. **5**(1), 17534 (2015)
32. Hendry, E., et al.: Chiral electromagnetic fields generated by arrays of Nanoslits. Nano Lett. **12**(7), 3640–3644 (2012)
33. Wiecha, P.R., et al.: Pushing the limits of optical information storage using deep learning. Nat. Nanotechnol. **14**(3), 237–244 (2019)
34. Meinders, E.R., et al.: Optical Data Storage: Phase-Change Media and Recording. Springer Science & Business Media (2006)
35. Gu, M., Li, X., Cao, Y.: Optical storage arrays: a perspective for future big data storage. Light Sci. Appl. **3**(5), e177–e177 (2014)
36. Van der Maaten, L., Hinton, G.: Visualizing data using t-SNE. J. Mach. Learn. Res. **9**(86), 2579–2605 (2008)

Chapter 6
Nanophotonic and Optical Platforms for Deep Learning

Abstract The preceding chapters introduced two classes of applications of DL in nanophotonics. In both cases, DL-based solutions provide an alternative to existing techniques, like advanced optimization algorithms for inverse design and electromagnetic solvers for forward modeling. These data-driven solutions, despite the pros and cons, are among the many advances ML has enabled across a wide spectrum of disciplines. From the standpoint of computer scientists, solving a problem from nanophotonics with DL may have nothing special in workflow compared with other tasks such as drug discovery or prediction of material properties; it is largely just a matter of feeding the ANNs data from another scientific domain. One of the most fascinating aspects of the interplay between DL and nanophotonics (or optics in general), which makes the latter a unique research field, is that the direction of application can be reversed. Owning the advantage in realizing massively parallel interconnections, nanophotonic systems as optical computing platforms have proved to be fast and power-efficient in implementing computational operations essential for ANNs. In the ideal case, once trained, they could run at the speed of light with little to no power consumption, leading to promising alternatives to electronic AI accelerators like GPU. In this chapter, we discuss photonic implementations of DL models for inference tasks (Wetzstein et al., Nature 588(7836):39–47, 2020). Foci will be given to all-optical nanophotonic circuits and free-space diffractive optics, each starting with showing how analogues can be established between computational operations and behaviors of light waves, followed by a representative proof-of-concept demonstration. Training of ANNs has different requirements on computation and thus will not be discussed in detail. Nonetheless, training in optical hardware is still possible in several ways (Wetzstein et al., Nature 588(7836):39–47, 2020; Wagner and Psaltis, Appl Opt 26(23):5061–5076, 1987; Hughes et al., Optica 5(7):864–871, 2018; Wright et al., Nature 601(7894):549–555, 2022). In addition to AI applications, optical and photonic systems have also been used in other paradigms such as neuromorphic computing (Shastri et al., Nat Photonics 15(2):102–114, 2021). The reader interested in these topics or in hybrid optical-electronic systems

---

The original version of this chapter was revised. The correction to this chapter is available at https://doi.org/10.1007/978-3-031-20473-9_7

K. Yao, Y. Zheng, *Nanophotonics and Machine Learning*, Springer Series in Optical Sciences 241, https://doi.org/10.1007/978-3-031-20473-9_6

for AI is referred to some recent review articles (Lima et al., Nanophotonics 6(3):577–599, 2017; Miller, J Lightwave Technol 35(3):346–396, 2017; Sande et al., Nanophotonics 6(3):561–576, 2017; Brunner et al., J Appl Phys 124(15):152004, 2018; Nahmias et al., IEEE J Sel Top Quant Electron 26(1):1–18, 2020; Zangeneh-Nejad et al., Nat Rev Mater 6(3):207–225, 2021; Wu et al., Engineering 10:133–145, 2022; Huang et al., Adv Phys X 7(1):1981155, 2022).

6.1 Deep Learning with Nanophotonic Circuits

Before diving into any specific nanophotonic implementations of ANNs, it is important to keep in mind that the recent development in this direction is a continuation of the earlier pursuit of analogue neural networks. As summarized in Fig. 6.1 taken from [1], the first optical demonstration of the Hopfield model, a type of recurrent neural networks, can be dated back to 1985 [2], even a year before the learning procedure of backpropagation was first described in the literature [3]. Although further advancement towards large-scale networks with reliable control over weights and nonlinearity was hindered in the 1990s by technical difficulties, with the considerable progress in integrated photonic circuits over the subsequent two decades, many of the major obstacles have been relieved or overcome. These improved capabilities stimulated research efforts to revisit the optical implementation of ANNs. In this section, we organize the discussion around the first coherent nanophotonic circuit for DL, reported by Shen et al. in 2017 [4], while another pioneering work of all-optical diffractive neural networks by Lin et al. [5] will be covered in the next section.

The applicability of optical and photonic systems to analogue computing for DL resides in the fact that the propagation of light through a medium or structure can be used to perform matrix multiplication. This led to the idea of universal linear optics [6–8], where a single programmable optical device, such as a photonic integrated

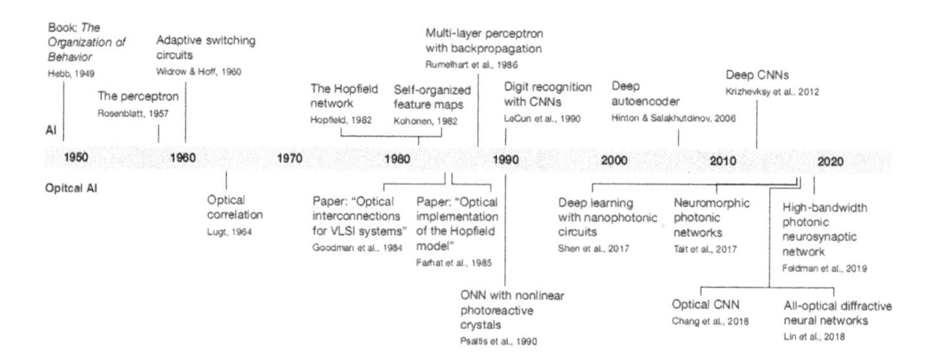

Fig. 6.1 Timeline of AI as well as its optical and photonic implementations. Selections of milestone works are made with preference to feature recent development. (Reprinted from [1] with permission of Springer Nature)

circuit (PIC) [9], can implement all possible linear transformations as linear operations of optical modes. To see how the analogue works, let us first take a quick look at some mathematics [10]. While an N-dimensional vector can be realized by N modes in a lossless optical system with N channels at the input and output each, the transformation between input and output modes is described by a unitary $N \times N$ matrix $U(N)$. It can be proven that an arbitrary unitary matrix U can be decomposed and represented by the product of a series of rotation matrices T_{mn} ($1 \leq m < n$)

$$U = DU' = D\left(\prod_{(m,n)\in S} T_{mn}\right), \tag{6.1}$$

with S defining a specific order of the multiplication,

$$T_{mn} = \begin{bmatrix} \begin{matrix} 1 & 0 \\ 0 & 1 \end{matrix} & \cdots & \cdots\ \cdots & \cdots & 0 \\ \vdots & \ddots & & & \vdots \\ \vdots & & \begin{matrix} t_{mm} & t_{mn} \\ t_{nm} & t_{nn} \end{matrix} & & \vdots \\ \vdots & & & \ddots & \vdots \\ 0 & \cdots & \cdots\ \cdots & \cdots & \begin{matrix} 1 & 0 \\ 0 & 1 \end{matrix} \end{bmatrix}, \tag{6.2}$$

and D a diagonal matrix with the modulus of each complex diagonal element being unity. The role of T_{mn} is to perform a transformation between channels m and n, which is a 2×2 unitary rotation, leaving the rest $N - 2$ dimensions of the matrix unchanged. The optical realization of the 2×2 mode transformation is usually based on beam splitters, which, in nanophotonics, are implemented by Mach-Zehnder interferometers (MZIs) [6, 9]. For a circuit with N channels, a total of $N(N - 1)/2$ MZIs are needed to allow connections between all the channels. Two popular arrangements proposed by Reck et al. [11] and Clements et al. [10], respectively, are sketched in Fig. 6.2a. Both of them feature mesh networks of beam splitters (shown as crosses), with the major difference being the sequence of crosses (i.e., S in Eq. (6.1)) that in turn determines the optical depth and compactness of the system. A second class of architectures known as recirculating meshes, which further enable routing light in the backward direction, is also widely used in programmable PICs; detailed discussions can be found in [9, 12, 13]. Figure 6.2b refines the mesh plots into layouts of waveguides and MZIs, and the diagonal matrix D is also included as phase shifters in each channel. In the typical configuration, an MZI consists of two 50% directional couplers and two phase shifters. One phase shifter of angle θ is set between the two directional couplers to control the split ratio of the MZI, and the other of angle φ is placed after the couplers to control the differential output phase. This gives rise to an SU(2) transformation between the input and output modes of the MZI:

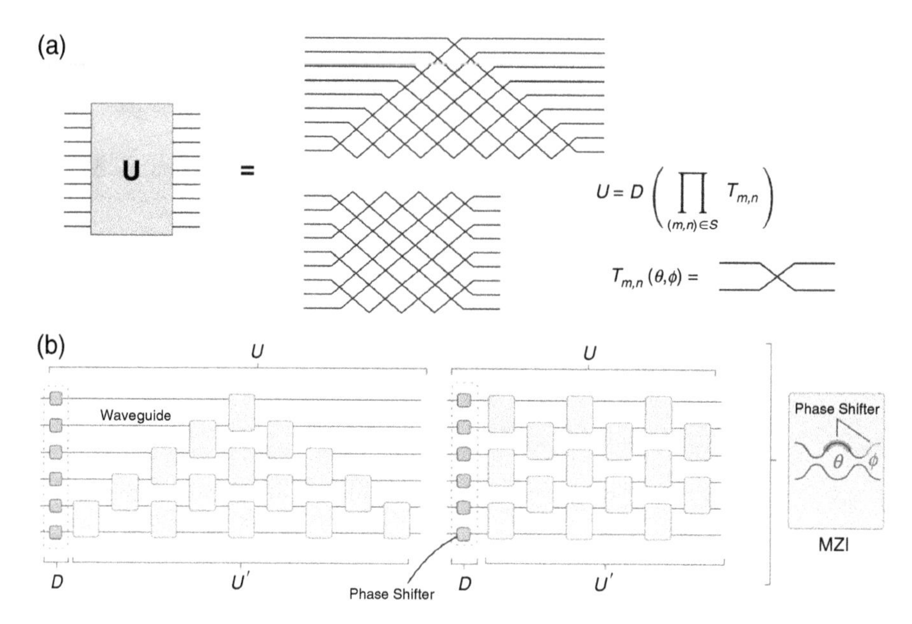

Fig. 6.2 (**a**) A universal unitary linear transformation U can be implemented by a mesh of beam splitters in the configuration proposed by Reck et al. (upper panel) or Clements et al. (lower panel). Each line represents an optical mode, and a crossing between two lines (channels m and n) denotes a beam splitter, described by a matrix T_{mn}. For a system with N channels, the mesh comprises $N(N-1)/2$ beam splitters. (**b**) Illustration of (**a**) in integrated photonic platforms. Each line corresponds to a waveguide. Beam splitters are realized by Mach-Zehnder interferometers (MZIs) denoted by gray boxes, each consisting of two 50% directional couplers and two phase shifters, with the internal one (θ) controlling the split ratio and the external one (φ) controlling the output phase, to perform 2 × 2 unitary rotation. The diagonal matrix D is realized by phase shifters on all channels, denoted by the purple squares. (Panel (b) is reprinted from [8] with permission. © 2018 Optica Publishing Group)

$$U(2) = \begin{pmatrix} e^{i\varphi}\cos\theta & -e^{i\varphi}\sin\theta \\ \sin\theta & \cos\theta \end{pmatrix}. \tag{6.3}$$

By replacing the elements t_{mm}, t_{mn}, t_{nm}, and t_{nn} in Eq. (6.2) with the elements as functions of θ and φ in Eq. (6.3), it becomes clearer why T_{mn} is introduced as a rotation matrix. Other variations of MZI blocks may differ slightly in configuration or involve more phase shifters, but they all function equivalently [14].

Many useful linear transformations are non-unitary. A general design framework which can tackle any linear operations was proposed by Miller [6]. The mathematical basis of this idea is that any linear transformation, when given in matrix form M, can be factorized using singular value decomposition (SVD) as [15]:

$$M = U\Sigma V^{\dagger}. \tag{6.4}$$

Here, U and V are $m \times m$ and $n \times n$ unitary matrices, respectively, with † denoting complex conjugate, and Σ is an $m \times n$ diagonal matrix with the diagonal elements (the singular values) usually chosen to be nonnegative real numbers [4, 8, 14]. The diagonal matrix Σ can be realized using different types of optical modulators for attenuation or amplification [4, 8]. Taking attenuators, for example, they can be implemented by MZIs, through which the transmissions of optical modes correspond to the singular values of Σ. Therefore, the circuit realization of a linear transformation M, according to Eq. (6.4), contains two blocks of MZIs for the universal unitary matrices U and $V^\dagger$, which are connected by a column of MZIs for the diagonal matrix Σ.

Instead of showing the layout of M as in Fig. 6.2 for individual unitary matrices, it is more informative to check out how the computational operations including matrix multiplications can be embodied in an optical version of ANNs. Figure 6.3a shows the design of an optical neural network (ONN) in the general architecture [4]. In addition to input and output being optical signals in photonic waveguides, each layer of the network is composed of an MZI-based optical interference unit (OIU)

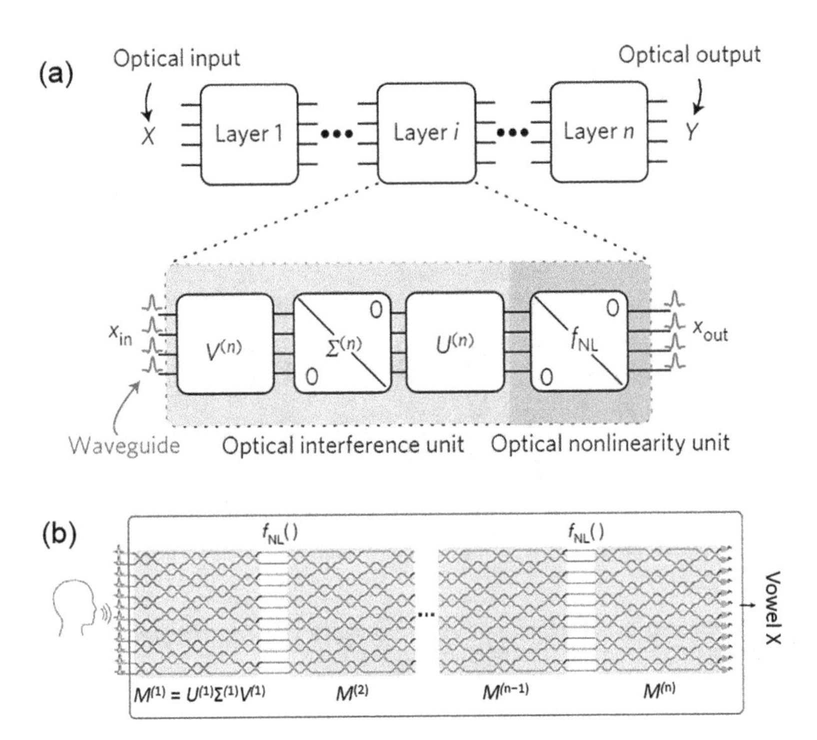

Fig. 6.3 (**a**) Top row: Optical implementation of an ANN as cascaded layers. Bottom row: Realization of an individual layer that comprises an optical interference unit (OIU) in the form of a universal linear network (gray block) and an optical nonlinearity unit (red block). Horizontal lines represent waveguides in the photonic circuit. Note that the layout of OIU is based on the factorization defined in Eq. (6.4). (**b**) Schematic of a photonic integrated circuit implementing an n-layer ANN for vowel recognition. (Reprinted from [4] with permission of Springer Nature)

implementing M in the decomposed form and an optical nonlinearity unit (ONU). In principle, with proper weighting factors controlled by the states of phase shifters in the MZIs, the PIC can perform inference tasks such as vowel recognition fully optically (Fig. 6.3b). Progress in parallel and afterwards also demonstrated different applications of quantum transport simulations [16] and complex-valued NNs for various classification tasks [17].

The actual realization of ONNs on photonic circuits, however, may still be limited by a few factors. One that could be fundamental is the scalability. As discussed above, an N-channel circuit needs $N(N-1)/2$ MZIs to build connections between all channels. This suggests that the number of MZIs grows as N^2 for N-dimensional data vectors. Although mass fabrication of circuit elements may no longer be a major challenge, many practical issues like losses and crosstalk will have non-negligible influence on circuit performance for large-scale networks [4]. In the experimental work by Shen et al., a 2-layer ONN employing 56 MZIs was used to demonstrate vowel recognition. The chip was not large enough to include the full network architecture all at once, so instead, a feedback and control loop was adopted to pass signals through the chip multiple times (Fig. 6.4). In each run, the OIU was reprogrammed to implement a part of the matrix multiplication ($U\Sigma$ or V) following

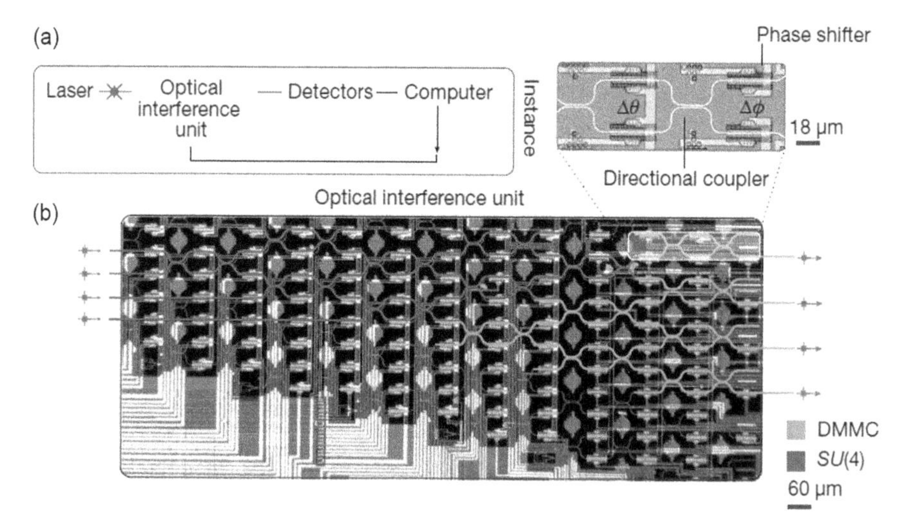

Fig. 6.4 Experimental demonstration of vowel recognition using a nanophotonic neural network. (**a**) Workflow of the experiments. Laser pulses encoding input data vectors are coupled to an OIU in a programmable circuit, measured by an array of photodiodes, and read on a computer in one instance. Unless inference is made at the output layer, processed signals are sent back to the OIU with reprogrammed weighting factors for the next instance. (**b**) Optical micrograph of the OIU used in the experiments. Meshes of MZIs implementing the universal unitary matrices, which are SU(4) transformations herein, are highlighted in red; those for attenuation or the diagonal matrix multiplication core (DMMC) are highlighted in blue. Layout of the MZIs follows the pattern described by Reck (see Fig. 6.2). Inset: A single MZI that has two directional couplers and four phase shifters, with the two internal (external) determining $\Delta\theta$ ($\Delta\varphi$). (Reprinted from [4] with permission of Springer Nature)

SVD. Another simplification was made to ONU. The realization of optical nonlinearity once suffered from efficiency problems at low signal intensities. Recent progress in saturable absorbers and other nonlinear thresholders has showed promise for more practical all-optical nonlinearity [1, 18], yet most experimental demonstrations thus far were still based on hybrid optical-electronic systems. With hybrid systems, optical signals need to be repeatedly measured, processed (electronically), and generated after each layer, as shown in Fig. 6.4a. For current challenges and advances in photonic implementations of nonlinearities, an insightful discussion can be found in a recent review [19]. In [4], nonlinearity was modeled on a computer as the response of saturable absorbers:

$$\sigma\tau_s I_{\text{in}} = \frac{1}{2}\frac{\ln\left(T_m / T_0\right)}{1 - T_m}. \tag{6.5}$$

Here, σ and τ_s are the absorption cross section and radiative lifetime of the absorbing material, I_{in} is the input intensity, T_0 is a constant initial transmittance, and T_m is the transmittance of the absorber to be solved for. The output intensity I_{out} is connected to the input by

$$I_{\text{out}} = I_{\text{in}} \cdot T_m\left(I_{\text{in}}\right), \tag{6.6}$$

which is depicted in Fig. 6.5.

The desired weighting factors stored in MZIs are determined as a result of training. Albeit several schemes for all-optical training of ANNs have been proposed, currently it is still more practical to conduct training on a computer. For the vowel recognition task in [4], 360 data points from 90 people speaking 4 different vowel phonemes were split to halves for training and testing, respectively. Since each sample must pass the chip twice for a complete matrix M due to the aforementioned

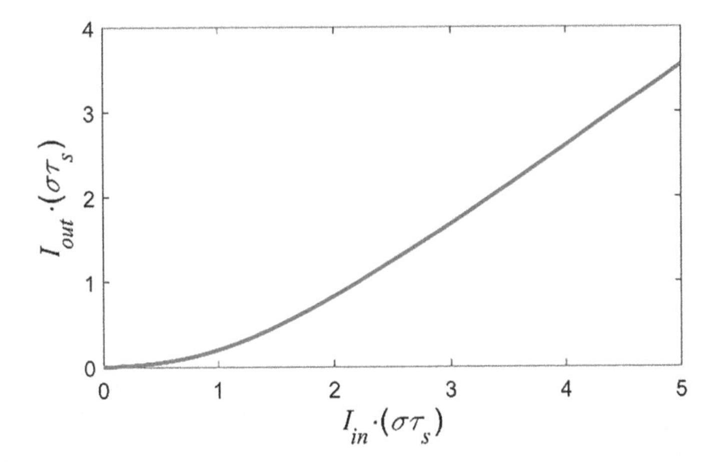

Fig. 6.5 Nonlinear response of a saturable absorber

reason, the inference over 180 test samples using the two-layer ONN required 720 experiments.

Figure 6.6a summarizes the experimental results for the ONN and a computer program implementing the same model. The ONN made 138 correct classifications out of 180 examples, giving an accuracy of 76.7%, somewhat worse than that of the electronic counterpart (91.7%). The difference in performance is largely attributed to the computational resolution of ONNs, which is limited by several factors related to the current hardware, such as photodetection noise, the precision of optical phase encoding, thermal crosstalk between phase shifters in the MZIs, etc. Another source of the misclassification, as seen in both systems for vowels C and D, comes from the proximity of these two vowels in the parameter space. In Fig. 6.6b, the projection of the test samples clearly reveals a larger overlap between vowels C and D. This is opposite to the situation of the example in Sect. 5.3. When samples cannot be separated into clusters, the classification becomes challenging, and there is a much larger chance that ANNs will make wrong decisions.

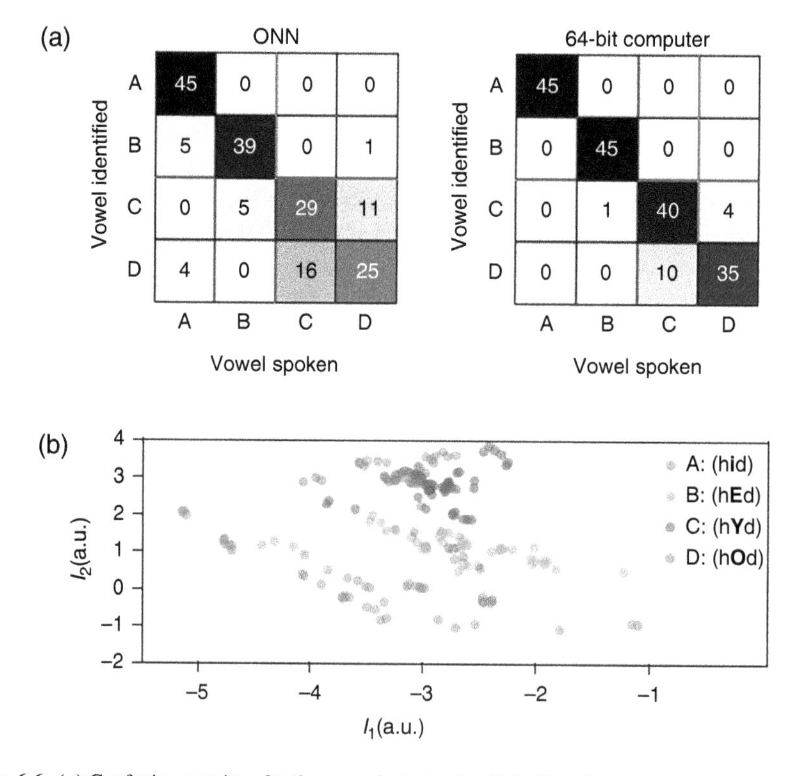

Fig. 6.6 (**a**) Confusion matrices for the nanophotonic circuit (left) and a computer program (right) implementing a two-layer neural network for vowel recognition. (**b**) 2D projection of the test dataset based on the log area ratio coefficients. The larger overlap between spoken vowels C and D (see legend) results in lower classification accuracy. (Reprinted from [4] with permission of Springer Nature)

6.2 All-Optical Diffractive Deep Neural Networks

A novel alternative to photonic circuits for implementing neural networks is based on free-space optics, and several demonstrations have been reported along this line [5, 20–26]. In order to adhere to the theme of this book, here we focus on the diffractive deep neural networks (D^2NNs) first proposed by Lin et al. [5]. For other schemes based on bulk optics and/or hybrid optical-electronic systems, interested readers are referred to recent reviews and references therein [1].

The theoretical basis of free-space implementations of neural networks is that many useful linear operations of waveforms naturally come with light propagation or light-matter interactions. A well-known example in Fourier optics is the Fourier-transform function of lenses. Similarly, correspondence can be established between linear matrix operations and light-matter interactions [1]. Figure 6.7a, b illustrates two elementary situations. In the first case, wave propagation in free space, according to the Fresnel-Kirchhoff diffraction formula, is described by a convolution of the wave with a complex-valued kernel. Because convolution is a basic computational operation in CNNs, this implies the utility of free-space optics in image-related DL applications, as long as the kernel can be engineered in some way. The second case involves a thin layer of scatterers. In a 2D space (Fig. 6.7b), when a light wave passes through the film, the corresponding operation is a multiplication with a diagonal matrix. More complicated transformations can be built upon these two elementary operations. For instance, cascaded layers of thin scatterers with spacing in between carry out successive multiplications of alternating diagonal matrices and convolution matrices, as illustrated in Fig. 6.7c.

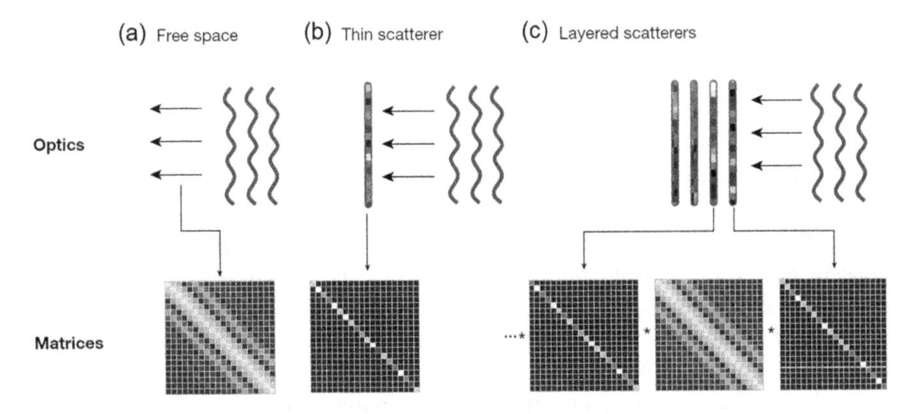

Fig. 6.7 Correspondence between wave propagation (top row) and linear matrix operations (bottom row). (**a**) Wave propagation in free space is mathematically related to a convolution of the field with a complex-valued kernel. (**b**) Wave propagation through a thin layer of scatterers can be described by a multiplication with a diagonal matrix. (**c**) Combining (**a**) and (**b**), the interaction between an incident wave and layered thin scatterers with spacing corresponds to successive diagonal matrices and convolution matrices. In (**b**) and (**c**), operations are counted from the first layer of scatterers. (Reprinted from [1] with permission of Springer Nature)

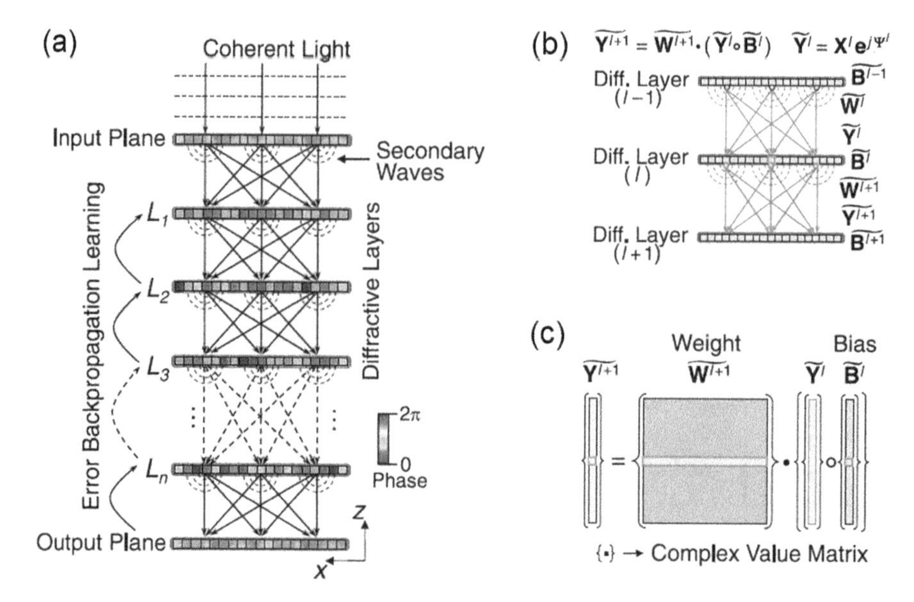

Fig. 6.8 Concept of diffractive deep neural networks (D^2NNs). (**a**) The analogy between coherent light passing through successive diffractive layers with spacing and an ANN. Similar to the interconnections between neurons in ANNs, each point in a layer is virtually connected to other points in the preceding layer through the interference of incoming secondary waves and its local transmission or reflection coefficient and to the points in the following layer through the secondary wave it generates and the local coefficients at those points. (**b**, **c**) Illustration of wave propagation in a D^2NN (**b**) and the corresponding matrix operations (**c**). Complex optical fields $\mathbf{Y}$ at a diffractive layer can be expressed with their magnitude $\mathbf{X}$ and phase $\mathbf{\Psi}$, which are modulated by the local transmission or reflection coefficient, mathematically as a multiplicative bias $\mathbf{B}$ via Hadamard product $\circ$. Further multiplication of the weighting matrices $\mathbf{W}$ is naturally performed by interlayer diffraction. (Reprinted from [5] with permission of AAAS)

The architecture of the all-optical D^2NN launched in [5] follows this configuration. The wave propagation process and analogue matrix operations are detailed in Fig. 6.8. Imagine that all the layers in the system, including the input and output planes and every intermediate diffractive layer, are pixelated. Then each pixel in the diffractive layers acts as a "neuron," and its connections to all the neurons in the adjacent layers are realized through propagation and interference of secondary waves. Unlike photonic circuit implementations, which are based on a more rigorous analogy to neural networks' architecture, a D^2NN possesses some unique differences in computation. The most important one is the function of individual neurons. In standard neural networks, a neuron applies a nonlinear activation function to the weighted sum of the outputs of the preceding layer, with all values involved being real and the weights and additive biases learnable. In contrast, data (encoded in waves) processed by a D^2NN are complex-valued, and the function of a neuron is to modulate the incoming waves by the local transmission or reflection coefficient, a bias term that is learnable but contributes via element-wise multiplication. A very interesting property of the D^2NN framework is, although nonlinearity can be incorporated to the system as in photonic circuits, the prototypes of D^{2}NN without inclusion of the equivalent of nonlinear activation functions still showed improved performance as the number of diffractive layers increased. In other words, multiple learnable diffractive layers appear to have more degrees of freedom than a single layer does on performing statistical inference and some other functions [26]. While there seems to be a contradiction between this property and the equivalence between a linear operation and its linear decomposition, possible explanations may hide behind practical factors, such as the finite sizes of the scatterers (not infinitesimally small) and of the layers (not infinitely large). A rigorous proof of the depth feature of D^{2}NNs will be beneficial to fully address the puzzle. To avoid causing confusion, it is probably more appropriate to term the present implementation as diffractive optical networks when nonlinearity is absent [27]. Despite this issue, the depth feature brings up a potential advantage of D^{2}NNs in reconfigurability: additional layers can be trained and added to an existing D^{2}NN for enhanced performance. It is also reasonable to expect the application of D^{2}NN framework to transfer learning [28, 29] and ensemble learning [30] due to their architectural compatibility.

Free-space-based all-optical neural networks can, in principle, be realized on different length scales. Since coherent light sources are readily available in almost all optical bands, the only constraint on implementing D^{2}NNs that can fall into the category of nanophotonic devices is the fabrication capability, including the creation of dense sub-micrometer-sized scatterers and precise alignment of the diffractive layers. The first experimental demonstration used a terahertz light source at 0.4 THz, corresponding to sizes of individual neurons spanning some hundreds of micrometers. Figure 6.9 shows two fabricated D^{2}NNs for classification and amplitude imaging, which consist of five and four layers of 3D-printed phase masks, respectively. For a given printing material and interlayer distances, the phase values at each neuron are determined from training on a computer and converted to height profiles. The operation of a device is simply done by inserting the input object and D^{2}NN into the optical path in front of a detector array. Compared to photonic integrated circuits, this realization of neural networks allows for a much larger number of neurons and thus learnable parameters. For instance, the five-layer device shown in Fig. 6.9 has 0.2 million neurons, in a scale comparable to the state-of-the-art DL models for sophisticated tasks such as image classification [5].

Fig. 6.9 3D-printed D^{2}NNs for classification (left) and amplitude imaging (right). (Courtesy of A. Ozcan)

The fabricated D²NN classifier was tested with images of handwritten digits taken from MNIST. Without any complex conversion, input images were encoded into optical fields by a mask at the input plane, as shown in Fig. 6.10a, b. The read-out of classification results was performed by an array of detectors at predefined locations on the output plane. As light propagation through and interference within the device lead to a redistribution of optical energy in space, only the detector corresponding to the correct label can receive a signal with maximum intensity. The performance of the fabricated D²NN and its electronic counterpart, based on tests with 50 and 10,000 samples, respectively, was summarized in Fig. 6.10c. The

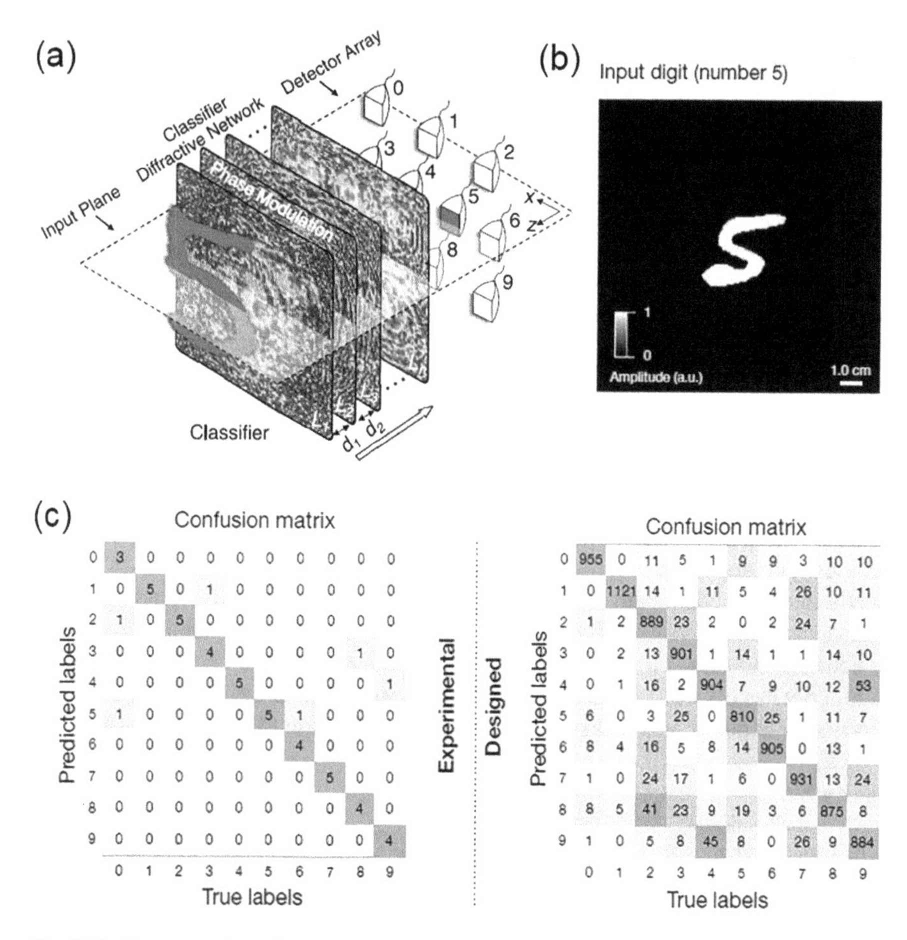

Fig. 6.10 Demonstration of all-optical classification of handwritten digits. (**a**) Schematic of the D²NN-based classifier. A 3D-printed digit as the target is placed in front of the device on the input plane. Under uniform illumination of terahertz waves, most of the transmitted energy is directed to the corresponding detector. Patterns on the layers of D²NN map the phase of transmission coefficient at each point. (**b**) An input image of digit 5. (**c**) Confusion matrices for the 3D-printed D²NN (left) and a computer program implementing the same model (right). (Reprinted from [1, 5] with permissions of AAAS and Springer Nature, respectively)

classification accuracy of the computer program was 91.75%, while the fabricated device got a match of 88%. In a simultaneous demonstration of classification of fashion products from MNIST, similar accuracies were obtained. These values confirm the validity of the D²NN framework, and further improvements towards the performance of state-of-the-art CNNs (~99.6%) could be expected with additional layers, integration of nonlinearities, and refined sample preparation.

6.3 In Situ Training of Photonic Neural Networks

The reader may have noticed that in the above two examples of nanophotonic circuits and D²NNs, training is not performed on the same optical device but elsewhere on a conventional computer. A fair concern about this protocol is, if training must be done electronically, not only are the programs based on idealized models of the real devices, but all the aforementioned advantages of optical platforms in speed and energy efficiency are limited to the final stage of ANN implementation. Albeit these "end products" still provide potential performance enhancement for many applications, having the training of photonic neural networks done in situ by optical means is attractive to a broader audience.

In the demonstration of photonic neural networks in [4], an on-chip training procedure without using backpropagation was discussed. The idea is that the error gradient for a specific weight parameter can be obtained with the "brute force" of finite difference method by sending the same batch of training data into the ONN twice, with the value of that weight parameter modified by a small change. This approach offers a possible solution to training simple ONNs, but because the gradients are calculated sequentially for each weight, it becomes very inefficient as the network scales up.

Hughes et al. proposed an alternative scheme for in situ training of photonic neural networks, which in theory is both efficient and scalable [31]. Essentially, they utilize the adjoint method to create a photonic analogue to the backpropagation algorithm. This, in the same way as how a photonic circuit performs inference tasks, allows the gradient to be determined in parallel at each weight element (i.e., phase shifters) through optical measurements. Figure 6.11a illustrates the operation of forward and backward propagations. The latter is adapted for the circuit implementation of OIUs where the weights are not given by matrix entries explicitly but parameterized as the permittivity of phase shifters. The loss gradient for a phase shifter in an arbitrary layer l then takes the form

$$\frac{d\mathcal{L}}{d\varepsilon_l} = \text{Re}\left\{\delta_l^T \frac{d\hat{W}_l}{d\varepsilon_l} X_{l-1}\right\}, \tag{6.7}$$

with X_{l-1} and δ_l^T denoting the input vector and transposed backpropagating error vector, respectively. Hughes et al. showed that the right side of Eq. (6.7) can be

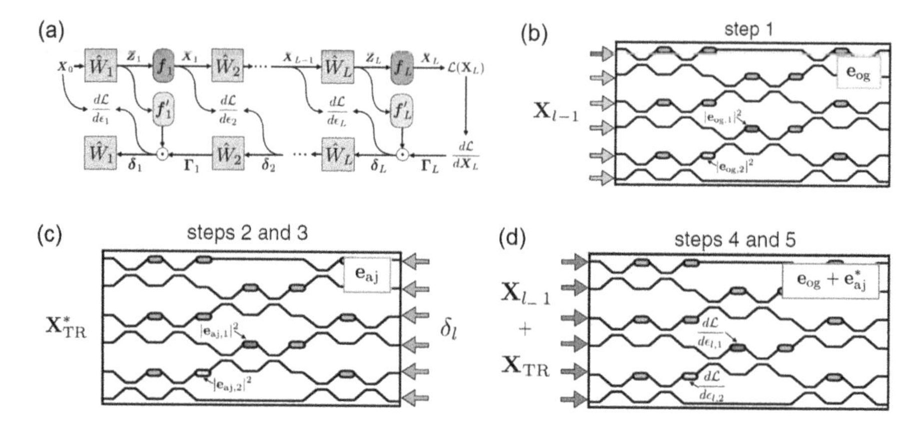

Fig. 6.11 In situ training of a photonic neural network. (**a**) Operation of forward/backward propagation (top/bottom row) and computation of the gradient of loss $\mathcal{L}$ with respect to the permittivity ε_l of each phase shifter. $\hat{W}_l$, transfer matrix of layer l relating inputs X_{l-1} and outputs Z_l. f_l, nonlinear activation function. The prime and ⊙ denote a derivative and element-wise multiplication, respectively. Γ_L, error vector, i.e., the difference between X_L and the ground truth. δ_l, derived vector of Γ_L as the error backpropagates. (**b–d**) Procedure for in situ measurements of the gradient at an OIU in layer l. The secondary subscript i ($i = 1, 2, \ldots$) denotes the i-th phase shifter. Red and yellow ovals highlight two tunable phase shifters for illustration purposes, among others in light blue. (**b**) Fields of the original signal X_{l-1} are sent in from the input ports. Intensities $|e_{og,i}|^2$ at each phase shifter are measured and stored. (**c**) Fields corresponding to the error δ_L are sent in from the output ports, resulting in adjoint fields e_{aj} in the OIU and X^*_{TR} at the input ports. Intensities $|e_{aj,i}|^2$ at each phase shifter are measured and stored. (**d**) Send $X_{l-1} + X_{TR}$ in from the input ports so that the original fields e_{og} and time-reversed adjoint fields e^*_{aj} interfere. The gradient of loss with respect to $\varepsilon_{l,i}$ can be obtained by measuring the intensities at each phase shifter and subtracting $|e_{og,i}|^2$ and $|e_{aj,i}|^2$ from them. (Reprinted from [31] with permission. © 2018 Optica Publishing Group)

expressed by the solutions to two electromagnetic problems defined on an OIU. In particular, X_{l-1} is related to the electric field distribution e_{og} in the circuit when excited on the input ports by signals of the training data, and δ_l^T is related to the field e_{aj} when excited on the output ports by sources corresponding to the loss. This is in line with the framework of adjoint method that involves an original (or direct) simulation and an adjoint simulation in each iteration, as introduced in Chap. 4. With some substitutions, Eq. (6.7) is rewritten to

$$\frac{d\mathcal{L}}{d\varepsilon_l} = k_0^2 \mathrm{Re}\left\{e_{aj} e_{og}\right\}. \tag{6.8}$$

Note that in the present context, the evaluation of the gradient takes place only at the positions of phase shifters. Equation (6.8) contains the fields from two propagation processes, and their spatial overlap is not directly measurable. Interestingly, this can be resolved by playing a trick to the adjoint field. Noting that the interference of e_{og} and the complex conjugate of e_{aj} leads to the following intensity pattern

$$I_{\left(e_{og}+e_{aj}^*\right)} = \left|e_{og}\right|^2 + \left|e_{aj}\right|^2 + 2\mathrm{Re}\left\{e_{aj} e_{og}\right\}, \tag{6.9}$$

the third term on the right side of our interest can be obtained by three intensity measurements on $\boldsymbol{e}_{og}$, $\boldsymbol{e}_{aj}$, and $e_{og} + e_{aj}^*$, respectively. Because the complex conjugate of a field corresponds to the time-reversed process, e_{aj}^* is physically generated by sources at the input ports with their amplitudes being the complex conjugate of what is received in the adjoint problem. The complete procedure is illustrated in Fig. 6.11b–d. Not limited to the numerical demonstration of an ONN-based XOR gate in [31], this in situ training protocol has recently been carried out experimentally on a hybrid digit-photonic platform for classification tasks [32]. The wide applicability could make it influential in adaptive optical and photonic networks [33] and quantum information processing [34].

References

1. Wetzstein, G., et al.: Inference in artificial intelligence with deep optics and photonics. Nature. **588**(7836), 39–47 (2020)
2. Farhat, N.H., et al.: Optical implementation of the Hopfield model. Appl. Opt. **24**(10), 1469–1475 (1985)
3. Rumelhart, D.E., Hinton, G.E., Williams, R.J.: Learning representations by back-propagating errors. Nature. **323**(6088), 533–536 (1986)
4. Shen, Y., et al.: Deep learning with coherent nanophotonic circuits. Nat. Photonics. **11**(7), 441–446 (2017)
5. Lin, X., et al.: All-optical machine learning using diffractive deep neural networks. Science. **361**(6406), 1004–1008 (2018)
6. Miller, D.A.B.: Self-configuring universal linear optical component. Photon. Res. **1**(1), 1–15 (2013)
7. Carolan, J., et al.: Universal linear optics. Science. **349**(6249), 711–716 (2015)
8. Harris, N.C., et al.: Linear programmable nanophotonic processors. Optica. **5**(12), 1623–1631 (2018)
9. Bogaerts, W., et al.: Programmable photonic circuits. Nature. **586**(7828), 207–216 (2020)
10. Clements, W.R., et al.: Optimal design for universal multiport interferometers. Optica. **3**(12), 1460–1465 (2016)
11. Reck, M., et al.: Experimental realization of any discrete unitary operator. Phys. Rev. Lett. **73**(1), 58–61 (1994)
12. Zhuang, L., et al.: Programmable photonic signal processor chip for radiofrequency applications. Optica. **2**(10), 854–859 (2015)
13. Pérez, D., et al.: Multipurpose silicon photonics signal processor core. Nat. Commun. **8**(1), 636 (2017)
14. Miller, D.A.B.: Perfect optics with imperfect components. Optica. **2**(8), 747–750 (2015)
15. Miller, D.A.B.: All linear optical devices are mode converters. Opt. Express. **20**(21), 23985–23993 (2012)
16. Harris, N.C., et al.: Quantum transport simulations in a programmable nanophotonic processor. Nat. Photonics. **11**(7), 447–452 (2017)
17. Zhang, H., et al.: An optical neural chip for implementing complex-valued neural network. Nat. Commun. **12**(1), 457 (2021)
18. Miscuglio, M., et al.: All-optical nonlinear activation function for photonic neural networks. Opt. Mater. Express. **8**(12), 3851–3863 (2018)
19. Shastri, B.J., et al.: Photonics for artificial intelligence and neuromorphic computing. Nat. Photonics. **15**(2), 102–114 (2021)

20. Chang, J., et al.: Hybrid optical-electronic convolutional neural networks with optimized diffractive optics for image classification. Sci. Rep. **8**(1), 12324 (2018)
21. Hughes, T.W., et al.: Wave physics as an analog recurrent neural network. Sci. Adv. **5**(12), eaay6946 (2019)
22. Yan, T., et al.: Fourier-space diffractive deep neural network. Phys. Rev. Lett. **123**(2), 023901 (2019)
23. Khoram, E., et al.: Nanophotonic media for artificial neural inference. Photon. Res. **7**(8), 823–827 (2019)
24. Wu, Z., et al.: Neuromorphic metasurface. Photon. Res. **8**(1), 46–50 (2020)
25. Qu, Y., et al.: Inverse design of an integrated-nanophotonics optical neural network. Sci. Bull. **65**(14), 1177–1183 (2020)
26. Mengu, D., et al.: Analysis of diffractive optical neural networks and their integration with electronic neural networks. IEEE J. Sel. Top. Quant. Electron. **26**(1), 1–14 (2020)
27. Mengu, D., et al.: Misalignment resilient diffractive optical networks. Nanophotonics. **9**(13), 4207–4219 (2020)
28. Qu, Y., et al.: Migrating knowledge between physical scenarios based on artificial neural networks. ACS Photonics. **6**(5), 1168–1174 (2019)
29. Veli, M., et al.: Terahertz pulse shaping using diffractive surfaces. Nat. Commun. **12**(1), 37 (2021)
30. Rahman, M.S.S., et al.: Ensemble learning of diffractive optical networks. Light Sci. Appl. **10**(1), 14 (2021)
31. Hughes, T.W., et al.: Training of photonic neural networks through in situ backpropagation and gradient measurement. Optica. **5**(7), 864–871 (2018)
32. Pai, S., et al.: Experimentally realized in situ backpropagation for deep learning in nanophotonic neural networks. arXiv preprint arXiv:2205.08501 (2022)
33. Pai, S., et al.: Parallel programming of an arbitrary feedforward photonic network. IEEE J. Sel. Top. Quant. Electron. **26**(5), 1–13 (2020)
34. Steinbrecher, G.R., et al.: Quantum optical neural networks. NPJ Quantum Inf. **5**(1), 60 (2019)

Correction to: Nanophotonics and Machine Learning

Kan Yao and Yuebing Zheng

Correction to: K. Yao, Y. Zheng, *Nanophotonics and Machine Learning*, Springer Series in Optical Sciences 241, https://doi.org/10.1007/978-3-031-20473-9

The original versions of Chapters 1, 2, 3, 4, and 6 were published inadvertently before all final corrections were made. Though updates were made to those chapters, the core idea was not changed.

The updated versions of the chapters can be found at:
https://doi.org/10.1007/978-3-031-20473-9_1
https://doi.org/10.1007/978-3-031-20473-9_2
https://doi.org/10.1007/978-3-031-20473-9_3
https://doi.org/10.1007/978-3-031-20473-9_4
https://doi.org/10.1007/978-3-031-20473-9_6

K. Yao, Y. Zheng, *Nanophotonics and Machine Learning*, Springer Series in Optical Sciences 241, https://doi.org/10.1007/978-3-031-20473-9_7

Index

K. Yao, Y. Zheng, *Nanophotonics and Machine Learning*, Springer Series in Optical Sciences 241, https://doi.org/10.1007/978-3-031-20473-9